CALCULUS
ESSENTIAL FOR STATISTICS

HUH, Myung-Hoe

Professor, Department of Statistics
Korea University

CHOI, Taeryon

Associate Professor, Department of Statistics
Korea University

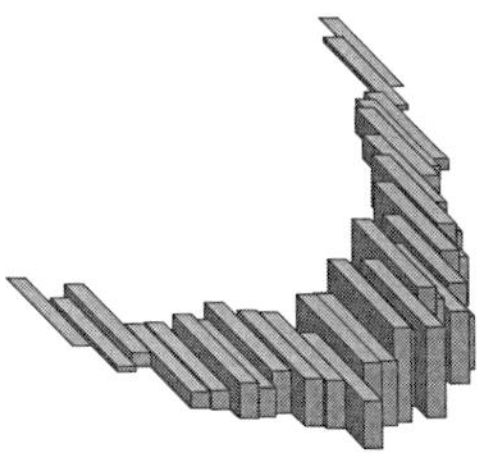

KYO WOO SA

KYO WOO SA
8, Yangnyeongsi-ro, Dongdaemun-gu,
Seoul 130-861 Korea

Phone: 82-2-925-2861
F a x: 82-2-925-2860
e-mail: kyowoo@kyowoo.co.kr
http://www.kyowoo.co.kr

ISBN 978-89-8172-945-5 (93410)
Printed in the Republic of korea.

Preface

Calculus is a fundamental tool in many areas of Science, and has an essential and fundamental role in the development of the basic probability theory underlying statistical methodology. Thus, knowledge of calculus has become indispensable for understanding the field of Statistics. A typical example where this occurs is in introductory probability or mathematical statistics, where students should have a working knowledge of elementary calculus, and the basic understanding of some materials that require summation, integration and differentiation, which are usually covered in calculus classes.

This book evolved from authors' lecture notes for one-semester core course on calculus for sophomore students in the department of Statistics at Korea University. Indeed, this book is a revised edition of "Lectures on Calculus for Statistics Students" by Prof. Myung-Hoe Huh, written in Korean. The Korean edition was intended to be at the same level with elementary calculus books such as *University Calculus* by Hass, Weir and Thomas or *Essential Calculus* by Stewart but provided a concise lecture notes for one-semester. Although there are many calculus textbooks, it is difficult for Statistics students with limited preparation to grasp the contents for one semester. For these reasons, there is a need for a basic calculus text which can be covered for a one-semester course, designed for undergraduate students who want to study Probability and Statistics.

In teaching the one-semester core course on calculus, our philosophy has been to try to expose students to as many of the the central ideas and topics of calculus essential for statistics. Thus, the objective of our course is to introduce calculus useful for Statistics and help students prepare for advanced courses, such as Probability and Mathematical Statistics. Therefore, in this book, we cover the core of calculus; differentiation and integration of functions of one and multi-variable with applications.

Most of the results and the proofs in this book undoubtedly resemble works found in other calculus books. Similarly, the suggested exercises at the ends of each chapter have come from many sources. One source used often was the file of old exams from STAT 201 that two authors have taught for several semesters. These problems, in turn, came from various sources which we could not keep every track of. In this regard, we apology for not having taken enough time to seek out the published reference and not giving proper credit. This book was typeset using TEX and figures were produced using R. We would like to thank our colleague, Professor Jayong Koo for helping us in editing the book with TEX.

Myung-Hoe Huh and Taeryon Choi, 2013

Contents

Chapter 1

Functions and Sequences

In this chapter, we begin by presenting some definitions, notations and properties that are basic to describing sequences and functions. These will be given in simplified form with a minimum of explanatory discussion.

1.1 Preliminaries

A **set** A will be defined as a "collection" (or "family") of "objects" called **elements** . The statement "x is an element of A" will be denoted as $x \in A$, and its negation will be denoted as $x \notin A$. The set with no elements will be called the **empty set** , and denoted as $\emptyset$.

Given a pair of sets, A and B, we say that A is a **subset** of B, and write $A \subset B$, if each element of A is an element of B. One obviously has $\emptyset \subset A$ for any set A. Two sets A and B are said to be **equal**, and we write $A = B$ if both $A \subset B$ and $B \subset A$.

If a set B contains at least one element that is not in A, then A is said to be a **proper subset** of B. The set that contains all elements under consideration in a certain study is called the **universal set** and is denoted as Ω or S.

A set is usually defined by either listing its elements explicitly, or by specifying a property that describes the elements of the set. If P denotes a property (statement) for elements of a set A, then we write

$$\{x \in A : P(x)\} \quad \text{or} \quad \{x \in A \mid \text{a statement}\}$$

for the set of all elements $x \in A$ for which the property P is true or that satisfy the statement.

Definition 1.1.1 (Set Operations) *Given two sets A and B, both subsets of a universal set Ω, we define their* **union** *to be the set*

$$A \cup B := \{x \in \Omega \,:\, x \in A \text{ or } x \in B\}.$$

Their **intersection** *is defined by*

$$A \cap B := \{x \in \Omega \,:\, x \in A \text{ and } x \in B\}.$$

We define the **complement** *of A (with respect to the universal set Ω) to be the set*

$$A^c := \{x \in \Omega \,:\, x \notin A\}.$$

We also define the **difference set** $A - B$ *by*

$$A - B := \{x \in A \,:\, x \notin B\} = A \cap B^c.$$

Definition 1.1.2 (The Real Number System)

- *The set of* **natural numbers** , $\mathbb{N} \triangleq \{1, 2, 3, \cdots\}$[1],
- *The set of* **integers** , $\mathbb{Z} \triangleq \{0, \pm 1, \pm 2, \ldots\}$,
- *The set of* **rational numbers** , $\mathbb{Q} := \{\frac{m}{n} \mid m, n \in \mathbb{Z} \;\; \textit{and}\; n \neq 0\}$,
- *The set of* **real numbers** , $\mathbb{R}$.
- *The set of* **irrational numbers** , $\mathbb{I} = \mathbb{R} - \mathbb{Q}$.
- $\mathbb{N} \subset \mathbb{Z} \subset \mathbb{Q} \subset \mathbb{R}$.

Unions and intersections can also be defined for families of sets. Suppose that we are given an **index set** Λ, and that, for each $\lambda \in \Lambda$, we are given a set $A_\lambda \subset \Omega$. We then define the union and intersection of the A_λ by

$$\bigcup_{\lambda \in \Lambda} A_\lambda := \{x \in \Omega \,:\, (\exists \lambda \in \Lambda)\ (x \in A_\lambda)\},$$

$$\bigcap_{\lambda \in \Lambda} A_\lambda := \{x \in \Omega \,:\, (\forall \lambda \in \Lambda)\ (x \in A_\lambda)\},$$

where the **universal quantifier** $\forall$ means "**for all**", and the **existential quantifier** $\exists$ means "**there exists**", or "**for some**".

[1]The symbol $:=$ and $\triangleq$ are used interchangeably to denote that the object on the left is being defined by the symbol on the right. However, we often use the symbol $=$ for all purposes.

Remark 1.1.1 (Quantifiers)

(1) The sentence "For every x, $x^2 - 3x + 2 = 0$" is a statement that can be classified as either true or false. In this case, it is a false statement. In symbols we write

$$\forall\ x,\ P(x),$$

where the **universal quantifier** $\forall$ is read as "for every...," "for all...", "for each...".

(2) The sentence "there exists an x such that $x^2 - 3x + 2 = 0$" is also a statement, and it is true. In symbols we write

$$\exists\ x\ s.t.\ P(x),$$

where the **existential quantifier** $\exists$ is read as "there exists...", "there is at least one...". The abbreviation **s.t.** is just a shorthand notation for the phrase "such that".

Example 1.1.1 (Interval Subsets of the Real Numbers)

- $[a, b]$ is called a **closed interval** , $[a, b] = \{x \in \mathbb{R} : a \leq x \leq b\}$.
- (a, b) is called an **open interval** , $(a, b) = \{x \in \mathbb{R} : a < x < b\}$.
- $[a, b)$ and $(a, b]$ are called **half-closed interval** or **half-open interval**, $[a, b) = \{x \in \mathbb{R} : a \leq x < b\}$, $(a, b] = \{x \in \mathbb{R} : a < x \leq b\}$.

Definition 1.1.3 *If A and B are sets, then the* **Cartesian product** *of A and B, written $A \times B$, is the set of all ordered pairs (a, b) such that*

$$A \times B = \{(a, b) : a \in A \text{ and } b \in B\}.$$

Definition 1.1.4 *Let A and B be sets. A* **function** *from A to B is a set $f \subset A \times B$ of ordered pairs in $A \times B$ such that for each $a \in A$ there exists a unique $b \in B$ with $(a, b) \in f$.*

In other words, the function from A to B is a rule which associates every element $x \in A$ to a unique element, denoted by $f(x)$, in B. If $(a, b) \in f$ and $(a, b') \in f$, then $b = b'$. The set A, the set of all first elements in f is called the **domain** of f, and the set B is referred to as the **codomain** of f. The set of all second elements in f is called the **range** of f, often denoted by $f(A) = \{f(x) \mid x \in A\}$. We also write

- $f : A \to B,\ \ f = \{(x, y)\ :\ x \in A,\ y = f(x)\}$,
- $f = \{(x, f(x))\ :\ x \in A\}$,
- $f(A) = \{y\ :\ y = f(x),\ x \in A\} \subset B$.
- The graph of a function $f : A \to B$ is $G_f = \{(x, f(x))\ |\, x \in A\}$.

Example 1.1.2 The unit circle, $x^2+y^2 = 1$ is not the graph of a function since it contains the graphs of two functions of x, $f(x) = \sqrt{1-x^2}$ and $f(x) = -\sqrt{1-x^2}$. That is, it does have two values $f(x)$ for some x, which violates the definition of a function.

Example 1.1.3 Sometimes a function is described by using different formulas on different parts of its domain.

(1) (**Integer floor** function) $y = \lfloor x \rfloor$ (Gauss function): greatest integer less than or equal to x.

(2) (**Integer ceiling** function) $y = \lceil x \rceil$: smallest integer greater than or equal to x.

(3) (Absolute value function) $y = |x|$.

There are a number of important types of functions frequently encountered in calculus.

- Power functions : A function $f(x)$ is a **power function** if $f(x) = x^a$, where a is a constant.

- Polynomial : A function $p(x)$ is a **polynomial** if

$$p(x) = a_n x^n + a_{n-1} x^{n-1} + \cdots + a_1 x + a_0,$$

  where n is a nonnegative integer and the numbers $a_0, a_1, \ldots, a_n$ are real constants, called the **coefficients** of the polynomial.

  - $p(x)$ is of order(degree) n if $a_n \neq 0$.
  - Polynomials of degree 1 are called **linear** functions.
  - Polynomials of degree 2 are called **quadratic** functions.
  - **Cubic** functions are polynomials of degree 3.

- Rational functions : A **rational** function is a quotient or ratio of two polynomials $p(x)$ and $q(x) \neq 0$,
$$f(x) = \frac{p(x)}{q(x)}.$$

- Algebraic functions : An **algebraic** function is a function constructed from polynomials using algebraic operations (addition, subtraction, multiplication, division and taking roots) of polynomials. Examples are
$$\frac{1}{x},\ 1+\frac{1}{x},\ \sqrt{x}\ (\text{ a square root}),\ \sqrt[3]{x}\ (\text{ a cube root}).$$

- Trigonometric functions : $f(x) = \sin x,\ \cos x,\ \tan x,\ \cdots.$

- Exponential functions : $f(x) = a^x > 0.$
The base a should be $a > 0,\ a \neq 1.$

- Logarithmic functions : $f(x) = \log_a x,\ x > 0$
The base a should be $a > 0,\ a \neq 1.$

- Functions that are not algebraic are called **transcendental** functions. Examples are trigonometric functions, exponential functions, logarithmic functions and so on.

Definition 1.1.5 (Even and Odd Functions) *A function $y = f(x)$ is an*

- **even** *function of x if $f(-x) = f(x)$. The graph of an even function is* **symmetric about the y-axis**.

- **odd** *function of x if $f(-x) = -f(x)$. The graph of an odd function is* **symmetric about the origin**.

Definition 1.1.6 (Increasing and Decreasing Functions) *A function $f : A \to \mathbb{R}$ is said to be* **increasing** *on A if whenever $x_1, x_2 \in A$ and $x_1 \leq x_2$, then $f(x_1) \leq f(x_2)$. (For* **decreasing** *, replace $f(x_1) \leq f(x_2)$ by $f(x_1) \geq f(x_2)$.) If a function is either increasing or decreasing on A, we say that it is* **monotone** *on A.*

Definition 1.1.7 (Properties of Functions)

- *A function $f : A \to B$ is called* **surjective** *or a map from A* **onto** *B if $f(A) = B$.*

- *A function $f : A \to B$ is called an* **injective** *(or a* **one-to-one** *) function, if for all a_1, a_2 in A, $f(a_1) = f(a_2)$ implies that $a_1 = a_2$. Equivalently, f is a one-to-one function, if whenever $a_1 \neq a_2$ one has $f(a_1) \neq f(a_2)$.*

- *A function $y = f(x)$ is one-to-one if and only if its graph intersects each horizontal line at most once.*

Definition 1.1.8 (Direct and Inverse Images). *Suppose that $f : A \to B$. If $C \subset A$, we let $f(C)$ represent the subset $\{f(x) : x \in C\}$ of B. The set $f(C)$ is called the* **(direct) image** *of C in B. For $D \subset B$, let $f^{-1}(D)$ represent the subset $\{x \in A : f(x) \in D\}$ of A. The set $f^{-1}(D)$ is called the* **inverse image** *(or* pre-image*) of D in A.*

Example 1.1.4 Suppose that a function $f : \mathbb{R} \to \mathbb{R}$ assigns $x \in \mathbb{R}$ to x^2. That is, $f(x) = x^2$. Then, the range of f is $f(\mathbb{R}) = [0, \infty)$, and there is no element that has an image less than 0. Thus, f is neither surjective nor injective. But restricting f to be defined on only $[0, \infty)$, it becomes injective. Thus, $f : [0, \infty) \to [0, \infty)$ is both surjective and injective, called **bijective** .

Definition 1.1.9 : Composite Function

When functions $f : A \to B$ and $g : B \to C$ are given, the **composite** *function $g \circ f : A \to C$ can be derived by the rule*

$$(g \circ f)(x) = g(f(x)), \quad for\ x \in A.$$

$g \circ f$ is read as "g composed with f".

Remark 1.1.2 The composition of two functions needs not be commutative. That is, $f \circ g \neq g \circ f$.

Example 1.1.5 Let $f(x) = \begin{cases} \frac{\pi}{x}, & \text{if } x \neq 0, \\ 0, & \text{if } x = 0. \end{cases}$, $g(y) = \sin y$. Then,

$$(g \circ f)(x) = \begin{cases} \sin \frac{\pi}{x}, & x \neq 0, \\ 0, & x = 0. \end{cases}$$

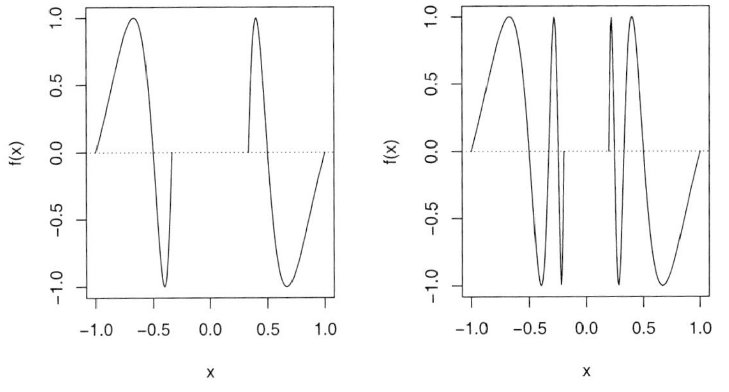

Figure 1.1: Graph of $\sin\frac{\pi}{x}$: [Left] $(-1,-1/3)$ and $(1/3,1)$, [Right]$(-1,-1/5)$ and $(1/5,1)$

Figure 1.1 shows the graph of $g \circ f$ between $(-1, 1)$. This function has the value 0 for all x satisfying $\pi/x = \pm k\pi$, $k = 1, 2, 3, \ldots$. In other words, if $x = \pm 1/k$, $k = 1, 2, 3, \ldots$, then $f(x) = 0$. Also, $f(x) = (-1)^k$ for x satisfying $\pi/x = \pm k\pi + \pi/2$. That is, if $x = 1/(\pm k + 0.5)$, $k = 1, 2, 3, \ldots$, then $f(x) = (-1)^k$. In addition, note that f is an odd function : $f(-x) = -f(x)$. The graph of $g \circ f$, the function oscillates between -1 and 1. Is f connected (or continuous) at $x = 0$ or not? Can you guess?

Definition 1.1.10 (Inverse function) *Let $f : A \to B$ be bijective (one-to-one and onto). The* **inverse function** *of f is the function f^{-1} given by*

$$f^{-1} = \{(y, x) \in B \times A : (x, y) \in f\}.$$

Remark 1.1.3 If $f : A \to B$ is one to one, then the inverse function of f, denoted by f^{-1} is a function which has domain $f(A)$ and range A. It satisfies: For $\forall y \in f(A)$,

$$f^{-1}(y) = x, \qquad where \qquad x \in A, \ \ f(x) = y.$$

When you refer f^{-1} as a function, f has to be one-to-one to make sure that f^{-1} is a function.

Example 1.1.6 The function $f(x) = \dfrac{2x}{x-1}$ is a bijection of $A = \{x \in \mathbb{R} : x \neq 1\}$ onto the set $B = \{y \in \mathbb{R} : y \neq 2\}$. the function inverse to f is given by

$$f^{-1}(y) = \frac{y}{y-2}, \quad \text{for} \;\; y \neq 2.$$

Example 1.1.7 Let $f : \mathbb{R} \to \mathbb{R}$, the inverse function of $f(x) = x^3$ is

$$f^{-1} : \mathbb{R} \to \mathbb{R},\; f^{-1}(y) = \begin{cases} y^{1/3}, & y \geqq 0, \\ -|y|^{1/3}, & y < 0. \end{cases}$$

Figure 1.2 provides the graphs of f and f^{-1}. We can see that two graphs are symmetric about $y = x$.

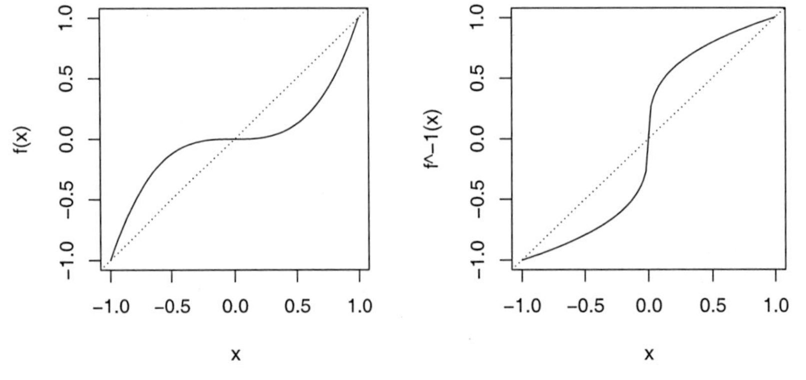

Figure 1.2: Graphs of f and f^{-1}

Theorem 1.1.1 *The graph of $f : A \to B$ and the graph of it's inverse function $f^{-1} : B \to A$ are symmetric about the line $y = x$.*

Proof Suppose that $(x_0, y_0) \in G_f$ (Here, G_f is the graph of f). In other words, $y_0 = f(x_0)$. Then, $x_0 = f^{-1}(y_0)$. Therefore, $(y_0, x_0) \in G_{f^{-1}}$. By the way, $P : (x_0, y_0)$ and $Q : (y_0, x_0)$ are located at symmetric position about the line $y = x$. Note that the center point of P and Q is $\left(\frac{x_0+y_0}{2}, \frac{y_0+x_0}{2}\right)$.

Theorem 1.1.2 *Let $f : A \to B$ and $g : B \to C$ be bijective. Then,*

(a) $(f^{-1} \circ f)(x) = x$ for all x in the domain of f.

(b) $(f \circ f^{-1})(y) = y$ for all y in the range of f.

(c) $(g \circ f)^{-1} = f^{-1} \circ g^{-1}$.

Proof Suppose that $y = f(x)$. Then

(a) $(f^{-1} \circ f)(x) = f^{-1}(f(x)) = f^{-1}(y) = x$, and

(b) $(f \circ f^{-1})(y) = f(f^{-1}(y)) = f(x) = y$.

(c) Suppose that $b = f(a)$ and $c = g(b)$. Then, $(g \circ f)(a) = c$. Thus,

$$(g \circ f)^{-1}(c) = a = f^{-1}(b) = f^{-1}(g^{-1}(c)) = (f^{-1} \circ g^{-1})(c).$$

Example 1.1.8 Suppose that $f : \mathbb{R} \to (0, 1)$, $f(x) = 2^x/(1 + 2^x)$. Find f^{-1}.

$$y = 2^x/(1 + 2^x) \ \Rightarrow \ \frac{y}{1-y} = 2^x \ \Rightarrow \ \log_2\left(\frac{y}{1-y}\right) = x.$$

Therefore, $f^{-1}(y) = \log_2 \frac{y}{1-y}$, for $0 < y < 1$. Also, we may write it as $f^{-1}(x) = \log_2 \frac{x}{1-x}$, for $0 < x < 1$.

Definition 1.1.11 *The inverse function of* $\sin x$ *is known as* $\arcsin x$. *Since $f(x) = \sin x$ is bijective on $-\pi/2 \le x \le \pi/2$, we can define its inverse function:*

$$y = \sin x \quad \Leftrightarrow \quad x = \sin^{-1} y \ \text{ for } -1 \le y \le 1, \ \ -\pi/2 \le x \le \pi/2.$$

1.2 Sequences

A sequence in the set A is a function of which the domain is the set $\mathbb{N}$ of natural numbers and of which the range is contained in A. In this section, we will discuss basic properties regarding sequences and the concept of the convergence of sequences.

Definition 1.2.1 *A* **sequence of real numbers** *(or a* **sequence in** $\mathbb{R}$*) is a function defined on the set $\mathbb{N} = \{1, 2, \ldots\}$ of natural numbers of which the range is contained in the set $\mathbb{R}$ of real numbers.*

If a is a sequence, we usually denote its value at n by a_n instead of $a(n)$. We call a_n the nth term of the sequence and often describe a sequence by giving a formula for the nth term.

Theorem 1.2.1 (Principle of Mathematical Induction) *Let $P(n)$ be a statement that is either true or false for each $n \in \mathbb{N}$. Then $P(n)$ is true for all $n \in \mathbb{N}$, provided that*

(1) $P(1)$ is true, and

(2) for each $k \in \mathbb{N}$, if $P(k)$ is true, then $P(k+1)$ is true.

Example 1.2.1 By induction, it is easy to show that

(1) $1 + 2 + 3 + \ldots + n = \frac{1}{2}n(n+1)$ for every natural number n.

(2) $7^n - 4^n$ is a multiple of 3, for all $n \in \mathbb{N}$.

Example 1.2.2 If $f_1(x) = x/(x+1)$ and $f_{n+1} = f_1 \circ f_n$ for $n = 1, 2, \ldots$, we can find a formula for $f_n(x)$ using the principle of mathematical induction.

Limit of Sequences

Definition 1.2.2 (Absolute Value) *For any real number $a \in \mathbb{R}$, we define the absolute value, denoted by $|a|$, to be :*

$$|a| := \begin{cases} a & \text{if } a \geq 0, \\ -a & \text{if } a < 0. \end{cases}$$

Theorem 1.2.2 (Triangle Inequality) *Given any $a, b \in \mathbb{R}$, we have the following:*

(1) $|a+b| \leq |a| + |b|$.

(2) $|a-b| \leq |a| + |b|$.

(3) $||a| - |b|| \leq |a-b|$.

Example 1.2.3 (Fibonacci Sequence) Fibonacci sequence is defined by $a_1 = 1$, $a_2 = 1$,

$$a_n = a_{n-1} + a_{n-2}, \text{ for } n = 3, 4, 5, \ldots.$$

For example, $a_3 = 2$, $a_4 = 3$, $a_5 = 5$, $\cdots$. according to this formula. What is the value of $\lim_{n\to\infty} a_n/a_{n-1}$? We need to define "the limit of a sequence" to answer this question.

Example 1.2.4 Consider the following sequences and determine their limits if they exist.

(a) $a_n = 1 + (-1)^n$.

(b) $a_n = 2n$.

(c) $a_n = 1 + 1/2^n$.

(d) $a_n = 1/n$.

(e) $a_n = (-1)^n$.

Definition 1.2.3 (Convergence, Limit). *Given a sequence $\{a_n\} \in \mathbb{R}$, we say that $\{a_n\}$* converges *to a real number a, and we write*

$$\lim a_n := \lim_{n\to\infty} a_n = a,$$

if, given any $\epsilon > 0$, there is an integer $N = N(\epsilon)$ such that $n \geq N$ implies $|a_n - a| < \epsilon$. The number a is called the limit *of the sequence $\{a_n\}$, and, if it exists, we say that the sequence is* **convergent** . *A sequence that has no limit is called* **divergent** .

Example 1.2.5 Let $a_n = 1 + \dfrac{1}{n}$. Then $\lim_{n\to\infty} a_n = 1$.

Note that for a given $\epsilon > 0$ we choose $N = [1/\epsilon] + 1$ (or $N > 1/\epsilon$). Then, $n > N$ implies that $|1 + \frac{1}{n} - 1| < \frac{1}{N} < \frac{1}{1/\epsilon} = \epsilon$.

Theorem 1.2.3 *Suppose that $\{a_n\}$ and $\{b_n\}$ are convergent sequences with $\lim a_n = a$ and $\lim b_n = b$. Then*

(1) $\lim_{n\to\infty}(k \cdot a_n) = ka$, *for $k \in \mathbb{R}$.*

(2) $\lim_{n\to\infty}(a_n \pm b_n) = a \pm b$.

(3) $\lim_{n\to\infty}(a_n \cdot b_n) = a \cdot b$.

(4) $\lim_{n\to\infty}(a_n/b_n) = a/b$, *provided that $b_n \neq 0$ for all n and $b \neq 0$.*

Proof Let us show that $\lim_{n\to\infty}(a_n + b_n) = a + b$. Since the theorem assumes the limits of sequences $\{a_n\}$ and $\{b_n\}$ for any $\epsilon > 0$, there exist $N_1 \in Z^+$ and $N_2 \in Z^+$ such that

$$n > N_1 \quad \Rightarrow \quad |a_n - a| < \epsilon/2,$$
$$n > N_2 \quad \Rightarrow \quad |b_n - b| < \epsilon/2.$$

Thus if we let $N = \max\{N_1, N_2\}$, then $n > N$ implies that

$$|(a_n + b_n) - (a + b)| \leq |a_n - a| + |b_n - b| < \epsilon/2 + \epsilon/2 = \epsilon.$$

Therefore, we conclude that

$$\lim_{n\to\infty}(a_n + b_n) = a + b.$$

Theorem 1.2.4 *If a sequence is convergent, its limit is unique. In other words, if $\{a_n\}$ converges to a and to a', then $a = a'$.*

Theorem 1.2.5 (Squeeze Theorem)

(1) Suppose that $\{a_n\}$ and $\{b_n\}$ are convergent sequences with $\lim_{n\to\infty} a_n = a$ and $\lim_{n\to\infty} b_n = b$. If $a_n \leq b_n$ for all $n \in \mathbb{N}$, then $a \leq b$.

(2) If $\{a_n\}$ converges to a and $a_n \geq 0$ for all $n \in \mathbb{N}$, then $a \geq 0$.

(3) Suppose that $\{a_n\}$, $\{b_n\}$, and $\{c_n\}$ are sequences of real numbers such that

$$a_n \leq c_n \leq b_n, \quad \forall n \in \mathbb{N}$$

and that $\lim_{n\to\infty} a_n = p = \lim_{n\to\infty} b_n$. Then $\{c_n\}$ is convergent and $\lim_{n\to\infty} c_n = p$.

Proof (3) Since $a_n - p \leq c_n - p \leq b_n - p$.

$$|c_n - p| \leq \max\{|a_n - p|, |b_n - p|\}.$$

By the way, since $\lim_{n\to\infty} a_n = \lim_{n\to\infty} b_n = p$, there exists $N \in Z^+$ such that $n > N$ implies $|a_n - p| < \epsilon$ and $|b_n - p| < \epsilon$, for any $\epsilon > 0$. Therefore,

$$n > N \Rightarrow |c_n - p| \leq \max\{|a_n - p|, |b_n - p|\} < \epsilon.$$

Example 1.2.6 Suppose that $c_n = \dfrac{\sin^3 n}{n^3}$. By taking $a_n = -\dfrac{1}{n^3}$ and $b_n = \dfrac{1}{n^3}$,

$$a_n \le c_n \le b_n, \quad \lim_{n\to\infty} a_n = 0 = \lim_{n\to\infty} b_n.$$

Thus, we have $\lim_{n\to\infty} c_n = 0$ by the squeeze theorem.

Theorem 1.2.6 *The following sequences converge to the limits listed below :*

(1) $\displaystyle\lim_{n\to\infty} \frac{\log n}{n} = 0.$

(2) $\displaystyle\lim_{n\to\infty} n^{1/n} = 1.$

(3) For $x > 0$, $\displaystyle\lim_{n\to\infty} x^{1/n} = 1.$

(4) For $|x| < 1$, $\displaystyle\lim_{n\to\infty} x^n = 0.$

(5) For any real number x, $\displaystyle\lim_{n\to\infty} \frac{|x|^n}{n!} = 0.$

(6) For any real number x, $\displaystyle\lim_{n\to\infty} \left(1+\frac{x}{n}\right)^n = e^x.$

Proof For (5), choose an integer $M > |x|$, so that $|x|/M < 1$. Then, $(|x|/M)^n \to 0$. Therefore for $n > N$,

$$\begin{aligned} \frac{|x|^n}{n!} &= \frac{|x|^n}{1\cdot 2\cdot\cdots\cdot M\cdot(M+1)(M+2)\cdot\cdots\cdot n} \\ &\le \frac{|x|^n}{M!M^{n-M}} = \frac{M^M}{M!}\left(\frac{|x|}{M}\right)^n. \end{aligned}$$

Note that the constant $M^M/M!$ does not change as n increases. Hence, the squeezing theorem tells us that $|x|^n/n! \to 0$.

Definition 1.2.4 (Increasing, Decreasing, Monotone, Bounded). *We say that a real sequence* $\{a_n\}$ *is* **increasing** *if* $a_n \le a_{n+1}$, *for all* $n \in \mathbb{N}$. *We say that it is* **decreasing** *if* $a_n \ge a_{n+1}$, *for all* $n \in \mathbb{N}$. *The sequence is* **monotone** *if it is either increasing or decreasing. Finally, the sequence is* **bounded** *(respectively,* **bounded above** *or* **bounded below** *) if the set* $\{a_n : n \in \mathbb{N}\}$ *is bounded (respectively, bounded above or bounded below). A sequence that is not bounded is called* **unbounded** .

Theorem 1.2.7 (Basic Properties)

- *The limit of a convergent sequence is unique.*
- *Every convergent sequence is bounded.*
- *Let $\{a_n\}$ be a real sequence. If $\{a_n\}$ is increasing and bounded above (or decreasing and bounded below), then $\{a_n\}$ is convergent.*

Definition 1.2.5 (Subsequence). *Let $\{a_n\}$ be a real sequence, and let $\{n_k\}$ be a sequence of natural numbers that is strictly increasing; i.e., $n_1 < n_2 < n_3 < \cdots$. Then the sequence $\{b_k\}$ defined by $b_k = a_{n_k} (= a_{n_1}, a_{n_2}, a_{n_3}, \ldots)$ is called a* **subsequence** *of $\{a_n\}$.*

Theorem 1.2.8 *Let $\{a_n\} \in \mathbb{R}$, with $\lim_{n\to\infty} a_n = a$. Then $\lim_{k\to\infty} a_{n_k} = a$, for any subsequence $\{a_{n_k}\}_{k\in\mathbb{N}}$. That is, every subsequence of $\{a_n\}$ also converges to the same limit a. However, the converse is not true.*

Theorem 1.2.9 *Let $a_n = \left(1 + \dfrac{1}{n}\right)^n$, $n = 1, 2, 3, \ldots$.*

(1) $\{a_n\}$ is an increasing sequence.

(2) $a_n \leq 3$.

(3) $\{a_n\}$ is a convergent sequence.

(4) We define the irrational number e to be the limit of a_n.

Remark 1.2.1 If we take a look at this sequence,

$$a_1 = 2,\ a_2 = (3/2)^2 = 2.25,\ a_3 = (4/3)^3 = 2.37,\ a_4 = (5/4)^4 = 2.44, \cdots$$

The limit of this sequence is the famous **Euler number** e, whose approximate value is $2.718281\cdots$, which is taken as the base of the "natural" logarithm. We can also define the number e in terms of infinite series :

$$e = \sum_{n=0}^{\infty} \frac{1}{n!}.$$

We can prove (1) and (2) of Theorem 1.2.9 using the following results.

Theorem 1.2.10 *Let* $0! = 1$ *and for* $n \in \mathbb{N}$ *define* $n!$ *by* $n \cdot (n-1)!$. *Let*

$$\binom{n}{r} = \frac{n!}{r!(n-r)!} \quad \text{for } r = 0, 1, 2, \ldots, n.$$

(1) Pascal's triangle :

$$\binom{n}{r-1} + \binom{n}{r} = \binom{n+1}{r}$$

for $r = 1, 2, \ldots, n$.

(2) Binomial theorem :

$$(a+b)^n = \sum_{r=0}^{n} \binom{n}{r} a^r b^{n-r},$$

for all numbers a *and* b *and positive integers* n.

(3) Bernoulli's inequality : If $1 + x > 0$, *then* $(1+x)^n \geq 1 + nx$ *for all* $n \in \mathbb{N}$.

(4) $2^{n-1} \leq n!$.

Proof of Theorem 1.2.9:

(1) By applying the Binomial theorem, we have

$$\begin{aligned} a_n &= \left(1 + \frac{1}{n}\right)^n = 1 + \binom{n}{1}\left(\frac{1}{n}\right)^1 + \binom{n}{2}\left(\frac{1}{n}\right)^2 + \cdots + \binom{n}{n}\left(\frac{1}{n}\right)^n \\ &= 1 + 1 + \frac{1}{2!}\left(1 - \frac{1}{n}\right) + \cdots + \frac{1}{n!}\left(1 - \frac{1}{n}\right)\cdots\left(1 - \frac{n-1}{n}\right). \end{aligned}$$

Similarly,

$$\begin{aligned} a_{n+1} &= \left(1 + \frac{1}{n+1}\right)^{n+1} \\ &= 1 + \binom{n+1}{1}\left(\frac{1}{n+1}\right)^1 + \binom{n+1}{2}\left(\frac{1}{n+1}\right)^2 + \cdots + \binom{n+1}{n}\left(\frac{1}{n+1}\right)^n \\ &\quad + \binom{n+1}{n+1}\left(\frac{1}{n+1}\right)^{n+1} \\ &= 1 + 1 + \frac{1}{2!}\left(1 - \frac{1}{n+1}\right) + \cdots + \frac{1}{n!}\left(1 - \frac{1}{n+1}\right)\cdots\left(1 - \frac{n-1}{n+1}\right) \\ &\quad + \frac{1}{(n+1)!}\left(1 - \frac{1}{n+1}\right)\cdots\left(1 - \frac{n}{n+1}\right) \\ &\geq 1 + 1 + \frac{1}{2!}\left(1 - \frac{1}{n+1}\right) + \cdots + \frac{1}{n!}\left(1 - \frac{1}{n+1}\right)\cdots\left(1 - \frac{n-1}{n+1}\right). \end{aligned}$$

Therefore,

$$\begin{aligned} a_{n+1}-a_n &= \left(1+\frac{1}{n+1}\right)^{n+1} - \left(1+\frac{1}{n}\right)^{n} \\ &\geq \left[1+1+\frac{1}{2!}\left(1-\frac{1}{n+1}\right)+\cdots+\frac{1}{n!}\left(1-\frac{1}{n+1}\right)\cdots\left(1-\frac{n-1}{n+1}\right)\right] \\ &\quad -\left[1+1+\frac{1}{2!}\left(1-\frac{1}{n}\right)+\cdots+\frac{1}{n!}\left(1-\frac{1}{n}\right)\cdots\left(1-\frac{n-1}{n}\right)\right] \\ &\geq 0+0+0+\cdots+0. \end{aligned}$$

(2) To show that a_n is bounded above, we note that $1/n! \leq 1/2^{n-1}$ by (4) of Theorem 1.2.10. Thus,

$$\begin{aligned} a_n &= 1+1+\frac{1}{2!}\left(1-\frac{1}{n}\right)+\cdots+\frac{1}{n!}\left(1-\frac{1}{n}\right)\cdots\left(1-\frac{n-1}{n}\right) \\ &\leq 1+1+\frac{1}{2!}+\cdots+\frac{1}{n!} \\ &\leq 1+1+\frac{1}{2^1}+\cdots+\frac{1}{2^{n-1}} \\ &\leq 1+\frac{1}{1-\frac{1}{2}} = 1+2. \end{aligned}$$

Therefore $a_n = \left(1+\frac{1}{n}\right)^n$ is an increasing sequence and bounded from above. Hence, the sequence a_n converges to a real number that is between 2 and 3. We denote the number e to be the limit of this sequence.

1.3 Limit of Functions

First, we state the precise definition of the limit of a function f at point c. Note that in this definition of limit, it does not matter whether f is defined at c or not.

Definition 1.3.1 *Let $f(x)$ be defined on an open interval about c except possibly at c itself. We say that the limit of $f(x)$ as x approaches c is a real number L and write $\lim_{x\to c} f(x) = L$, if for each (given any) $\epsilon > 0$ there exists a $\delta > 0$ such that $|f(x)-L| < \epsilon$ whenever $0 < |x-c| < \delta$. Equivalently, we say that the limit of $f(x)$ as x approaches c is L if there exists some $\delta > 0$ such that $0 < |x-c| < \delta$ implies $|f(x)-L| < \epsilon$ for any $\epsilon > 0$.*

Remark 1.3.1 The real number L is said to be the **limit of f at c** . Since the value of δ usually depends on ϵ, we may write $\delta(\epsilon)$ to emphasize this dependence. If the limit of f at c does not exist, we say that f **diverges** at c.

Example 1.3.1 We give some examples to illustrate how the precise definition of the limit is applied.

- $\lim_{x\to c} b = b$.
- $\lim_{x\to c} x = c$.
- $\lim_{x\to c} 3x + 5 = 3c + 5$.
- $\lim_{x\to c} x^2 = c^2$.

Remark 1.3.2 (Sequential criterion)

- Let $A \subset \mathbb{R}$. A point c in $\mathbb{R}$ is an **accumulation point** or a **cluster point** of A if every $\epsilon > 0$ there exists at least one point $x \in A$, $x \neq c$ such that $|x - c| < \epsilon$.
- A number $c \in \mathbb{R}$ is a cluster point of a subset A of $\mathbb{R}$ if and only if there exists a sequence $\{a_n\}$ in A such that $\lim a_n = c$ and $a_n \neq c$ for all $n \in \mathbb{N}$.
- If $f : A \to \mathbb{R}$ and if c is a cluster point of A, then f can have only one limit at c.
- **(Sequential criterion)** Let $f : A \to \mathbb{R}$ and let c be a cluster point of A. Then the following statements are equivalent.

  (a) $\lim\limits_{x\to c} f(x) = L$.

  (b) For every sequence $\{a_n\}$ in A that converges to c such that $a_n \neq c$ for all $n \in \mathbb{N}$, the sequence $\{f(a_n)\}$ converges to L.
- **(Divergence criterion)** The function f does **not** have a limit at c if and only if there exists a sequence $\{a_n\}$ in A with $a_n \neq c$ for all $n \in \mathbb{N}$ such that the sequence $\{a_n\}$ converges to c but the sequence $\{f(a_n)\}$ does **not** converge in $\mathbb{R}$.

Example 1.3.2 Consider the following functions and their limits.

(1) $\lim\limits_{x\to 0} \dfrac{1}{x}$ does not exist.

(2) $\lim_{x \to 0} \operatorname{sgn}(x)$ does not exist.

(3) $\lim_{x \to 0} \dfrac{x}{|x|}$ does not exist.

(4) $\lim_{x \to 0} \sin(1/x)$ does not exist.

Theorem 1.3.1 (Limit Theorems) *Let $A \subset \mathbb{R}$, let f and g be functions on A to $\mathbb{R}$, and let $c \in \mathbb{R}$ be a cluster point of A. If $\lim_{x \to c} f(x) = L$ and $\lim_{x \to c} g(x) = M$, then*

(1) $\lim_{x \to c} \{f(x) \pm g(x)\} = L \pm M$

(2) $\lim_{x \to c} \{f(x) \cdot g(x)\} = L \cdot M$

(3) $\lim_{x \to c} \{k \cdot f(x)\} = k \cdot L,\ k \in \mathbb{R}.$

(4) $\lim_{x \to c} \dfrac{f(x)}{g(x)} = \dfrac{L}{M},\ M \neq 0$

Proof (1) Since $\lim_{n \to c} f(x) = L$ and $\lim_{n \to c} g(x) = M$, there exist $\delta_1 > 0$ and $\delta_2 > 0$ such that

$$0 < |x - c| < \delta_1 \textit{ and } 0 < |x - c| < \delta_2$$

$$\Rightarrow\ |f(x) - L| < \epsilon_1 (= \epsilon/2) \textit{ and } |g(x) - M| < \epsilon_2 (= \epsilon/2)$$

for every $\epsilon > 0$. Hence

$$0 < |x - c| < \min(\delta_1, \delta_2)(= \delta)$$

$$\Rightarrow\ |f(x) + g(x) - L - M| \leq |f(x) - L| + |g(x) - M| < \epsilon_1 + \epsilon_2 = \epsilon.$$

Therefore, $\lim_{n \to c} \{f(x) + g(x)\} = L + M$.

Example 1.3.3 Some of the limits can be easily established :

(1) $\lim_{x \to c} \dfrac{1}{x} = \dfrac{1}{\lim_{x \to c} x} = \dfrac{1}{c}.$

(2) $\lim_{x \to 1} (x^2 + 2x) = 3.$

(3) $\lim_{x \to 2} \dfrac{x^3 - 2}{x^2 + 2} = 1.$

(4) $\lim_{x \to 2} \dfrac{x^2 - 4}{5x - 10} = \dfrac{4}{5}.$

(5) If $p(x)$ is a polynomial function, then $\lim_{x \to c} p(x) = p(c)$.

(6) If $p(x)$ and $q(x)$ are polynomial functions and if $q(c) \neq 0$, then

$$\lim_{x \to c} \frac{p(x)}{q(x)} = \frac{\lim_{x \to c} p(x)}{\lim_{x \to c} q(x)} = \frac{p(c)}{q(c)}.$$

Theorem 1.3.2 (Squeeze Theorem) *Suppose that $g(x) \leq f(x) \leq h(x)$ for all x in some open interval containing c, except possibly at $x = c$ itself. Suppose also that*

$$\lim_{x \to c} g(x) = \lim_{x \to c} h(x) = L.$$

Then $\lim_{x \to c} f(x) = L$.

Proof Since $\lim_{x \to c} g(x) = L$ and $\lim_{x \to c} h(x) = L$, there exist $\delta_1 > 0$ and $\delta_2 > 0$ for every $\epsilon > 0$ such that

$$\begin{aligned} & 0 < |x - c| < \delta_1 \text{ and } 0 < |x - c| < \delta_2 \\ \Rightarrow \quad & -\epsilon < g(x) - L, \; h(x) - L < \epsilon, \\ \textit{or} \quad & -\epsilon < g(x) - L \leq f(x) - L \leq h(x) - L < \epsilon. \end{aligned}$$

Take $\delta = \min(\delta_1, \delta_2)$. Then, for a given $\epsilon > 0$, $|x - c| < \delta$ implies that $|f(x) - L| < \epsilon$.

Theorem 1.3.3 *If $f(x) \leq g(x)$ for all x in some open interval containing c, except possible at $x = c$ itself, and the limits of f and g both exist as x approaches c, then*

$$\lim_{x \to c} f(x) \leq \lim_{x \to c} g(x)$$

Example 1.3.4 Some applications of the Squeeze Theorem are :

(1) $\lim_{x \to 0} |x|^{3/2} = 0.$

(2) $\lim_{x \to 0} x \sin(1/x) = 0.$

(3) $\lim_{x \to 0} \sin x = 0.$

(4) $\lim_{x \to 0} \dfrac{\sin x}{x} = 1.$

(5) $\lim_{x \to 0} \dfrac{\cos x - 1}{x} = 0.$

In order to prove (3) and (4), we use a geometric argument. Assume first that x lies between 0 and $\pi/2$. Then consider a sector of a circle with center O, central angle x and radius 1, as in Figure 1.3. A sector OAB is made by the arc connecting point A and B on the circumference and the point O which is a center of the circle. Note that $\overline{OA} = 1$. Since x is the angle between two sides at the center O, the area of sector OAB is $x/2$ and the sector OAB is bigger than the triangle $\triangle OAB$ inscribed in the sector.

Thus, we have $0 < \sin x < x$, since

$$0 < \frac{\sin x}{2} = \text{ Triangle } OAB < \text{ Sector } OAB = \frac{x}{2}. \tag{1.3.1}$$

For $-\pi/2 < x < 0$, $0 < \sin(-x) < -x$ or $0 < -\sin x < -x$. Hence

$$|\sin x| < |x|, \ \forall\, x \text{ such that } 0 < |x| < \tfrac{\pi}{2}. \tag{1.3.2}$$

By the Squeeze Theorem, we have $\lim_{x \to 0} \sin x = 0$.

Also for $0 < x < \pi/2$, note that the sector OAB is smaller than $\triangle OAC$ that contains the sector OAB as shown in Figure 1.3.

$$\frac{\sin x}{2} = \ \triangle\, OAB < \text{ Sector } OAB = \frac{x}{2} < \triangle\, OAC = \frac{\tan x}{2}. \tag{1.3.3}$$

Hence, by taking the reciprocals, we have

$$\frac{1}{\sin x} > \frac{1}{x} > \frac{\cos x}{\sin x}. \tag{1.3.4}$$

Therefore, multiplying $\sin x > 0$ to (1.3.4), we have

$$1 > \frac{\sin x}{x} > \cos x.$$

By the Squeeze Theorem, we have

$$\lim_{x \to 0+} \frac{\sin x}{x} = 1.$$

Furthermore, note that $f(x) = \dfrac{\sin x}{x}$ is an even function, which concludes that

$$\lim_{x \to 0} \frac{\sin x}{x} = 1.$$

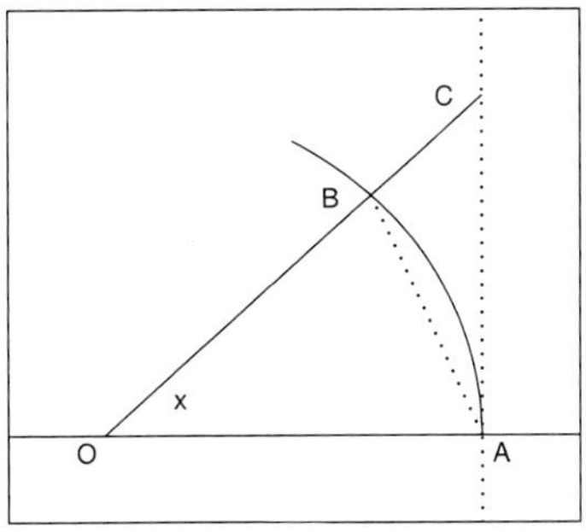

Figure 1.3: Sector OAB and triangle OAB, and triangle OAC.

Definition 1.3.2 (One-sided Limits)

Let $A \subset \mathbb{R}$ and let $f : A \to \mathbb{R}$.

(1) We say that $L \in \mathbb{R}$ is the **right-hand limit of** *f at c and we write*

$$\lim_{x \to c^+} f(x) = L,$$

if given any $\epsilon > 0$ there exists a $\delta = \delta(\epsilon) > 0$ such that, for all $x \in A$ with $0 < x - c < \delta$, $|f(x) - L| < \epsilon$.

(2) We say that $L \in \mathbb{R}$ is the **left-hand limit of** *f at c and we write*

$$\lim_{x \to c^-} f(x) = L,$$

if given any $\epsilon > 0$ there exists a $\delta = \delta(\epsilon) > 0$ such that, for all $x \in A$ with $-\delta < x - c < 0$, $|f(x) - L| < \epsilon$.

Remark 1.3.3 $x \to c^+$ can be written as $x \downarrow c$ or $x \to c,\ x > c$, and $x \to c^-$ can be written as $x \uparrow c$ or $x \to c,\ x < c$.

Theorem 1.3.4 *Let $A \subset \mathbb{R}$, let $f : A \to \mathbb{R}$, and let $c \in \mathbb{R}$ be a cluster point of both of the sets $A \cap (c, \infty)$ and $A \cap (-\infty, c)$. Then $\lim_{x\to c} f(x) = L$ if and only if $\lim_{x\to c^+} f(x) = L = \lim_{x\to c^-} f(x)$.*

Example 1.3.5 Consider the following functions and their limits.

(a) $f(x) = \mathrm{sgn}(x)$. Then, $\mathrm{sgn}(x)$ does not have a limit at 0.

(b) $f(x) = \lfloor x \rfloor$. Then, $f(x)$ does not have a limit at all integers.

(c) $f(x) = e^{1/x}$ for $x \neq 0$. Then, $\lim\limits_{x\to 0+} e^{1/x}$ does not exist in $\mathbb{R}$ while $\lim\limits_{x\to 0-} e^{1/x} = 0$. Thus, the limit at $x = 0$ does not exist.

Definition 1.3.3 (Infinite Limits) *Let $A \subset \mathbb{R}$, let $f : A \to \mathbb{R}$.*

(1) We say that f **tends to** ∞ *as $x \to c$, and write*

$$\lim_{x\to c} f(x) = \infty,$$

if, for every $\alpha \in \mathbb{R}$, there exists $\delta = \delta(\alpha) > 0$ such that, for all $x \in A$ with $0 < |x - c| < \delta$, $f(x) > \alpha$.

(2) We say that f **tends to** $-\infty$ *as $x \to c$, and write*

$$\lim_{x\to c} f(x) = -\infty,$$

if, for every $\beta \in \mathbb{R}$, there exists $\delta = \delta(\beta) > 0$ such that, for all $x \in A$ with $0 < |x - c| < \delta$, $f(x) < \beta$.

Definition 1.3.4 (Limits at Infinity) *Let $A \subset \mathbb{R}$, let $f : A \to \mathbb{R}$. Suppose that $(a, \infty) \subset A$ for some $a \in \mathbb{R}$. We say that $L \in \mathbb{R}$ is the* **limit of** f *as $x \to \infty$, and write*

$$\lim_{x\to\infty} f(x) = L,$$

if, given any $\epsilon > 0$, there exists $M = M(\epsilon) > a$ such that, for any $x > M$, $|f(x) - L| < \epsilon$.

Theorem 1.3.5 *Let $A \subset \mathbb{R}$, let $f : A \to \mathbb{R}$, and suppose that $(a, \infty) \subset A$ for some $a \in \mathbb{R}$. Then the followings are equivalent.*

(1) $\lim_{x\to\infty} f(x) = L$

(2) For every sequence $\{x_n\}$ *in* $A \cap (a, \infty)$ *such that* $\lim_{n\to\infty} x_n = \infty$, *the sequence* $\{f(x_n)\}$ *converges to* L.

Example 1.3.6 (1) $\displaystyle\lim_{x\to 0} \frac{1}{x^2} = \infty$.

(2) $\displaystyle\lim_{x\to 0} \frac{1}{x}$ does not exist.

(3) $\displaystyle\lim_{x\to\infty} \frac{1}{x} = 0$.

(4) $\displaystyle\lim_{x\to-\infty} x^n = \infty$ for even n, and $\displaystyle\lim_{x\to-\infty} x^n = -\infty$ for odd n.

(5) $\displaystyle\lim_{x\to\infty} e^{-x} = 0$.

Definition 1.3.5 (Asymptotes) *A line* $x = a$ *is a vertical asymptote of the graph of a function* $y = f(x)$ *if either*

$$\lim_{x\to a^+} f(x) = \pm\infty \quad or \quad \lim_{x\to a^-} f(x) = \pm\infty.$$

A line $y = b$ *is a horizontal asymptote of the graph of a function* $y = f(x)$ *if either*

$$\lim_{x\to\infty} f(x) = b \quad or \quad \lim_{x\to-\infty} f(x) = b.$$

Example 1.3.7 We can find the asymptotes of the curve.

(1) For $f(x) = \dfrac{x+1}{x+2}$, $x = -2$ and $y = 1$. See Figure 1.4.

(2) For $f(x) = \sqrt{x^2+2} - x$, $y = 0$.

(3) For $f(x) = \tan x$, $x = (n + 1/2)\pi$, where n is an integer.

Example 1.3.8 The followings are useful facts regarding e and e^x :

(1) $\displaystyle\lim_{n\to\infty} \left(1 + \frac{1}{n}\right)^n = e$

(2) $\displaystyle\lim_{x\to 0} (1+x)^{\frac{1}{x}} = e$.

We give a brief argument for (2).

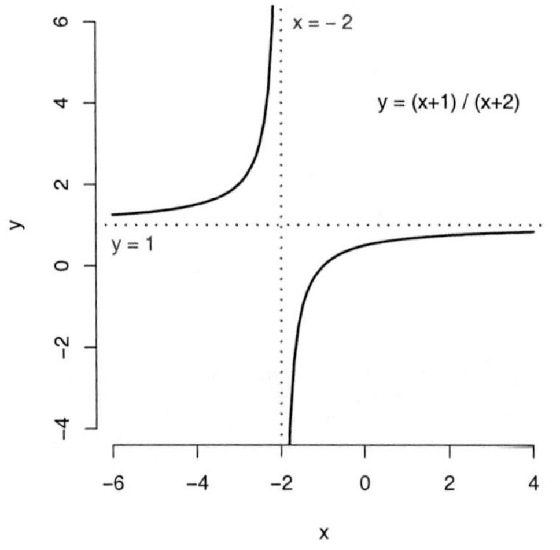

Figure 1.4: The lines $y = 1$ and $x = -2$ are asymptotes of the curve (1) in Example 1.3.7.

(a) For $x > 0$:

Since $\left\lfloor \dfrac{1}{x} \right\rfloor \le \dfrac{1}{x} < \left\lfloor \dfrac{1}{x} \right\rfloor + 1$ or $\dfrac{1}{\lfloor 1/x \rfloor} \ge x > \dfrac{1}{\lfloor 1/x \rfloor + 1}$, we have

$$(1+x)^{\lfloor 1/x \rfloor} \le (1+x)^{1/x} < (1+x)^{\lfloor 1/x \rfloor + 1}.$$

Therefore,

$$\left(1 + \frac{1}{\lfloor 1/x \rfloor + 1}\right)^{\lfloor 1/x \rfloor} < (1+x)^{1/x} < \left(1 + \frac{1}{\lfloor 1/x \rfloor}\right)^{\lfloor 1/x \rfloor + 1}.$$

By the way, since

$$\lim_{x \to 0+} \left(1 + \frac{1}{[1/x] + 1}\right)^{[1/x]} = e, \quad \lim_{x \to 0+} \left(1 + \frac{1}{[1/x]}\right)^{[1/x]+1} = e,$$

we have $\lim_{x \to 0+} (1+x)^{1/x} = e$ by the Squeeze Theorem.

(b) For $x < 0$:

Let $y = -x$, $z = \frac{y}{1-y}$.

$$\lim_{x \to 0^-} (1+x)^{\frac{1}{x}} = \lim_{y \to 0+} \left(\frac{1}{1-y}\right)^{\frac{1}{y}} = \lim_{z \to 0+} (1+z)^{1+\frac{1}{z}} = e.$$

(3) $\lim_{x \to 0} \frac{1}{x} \log(1+x) = 1.$

(4) $\lim_{x \to 0} \frac{e^x - 1}{x} = 1.$

(5) $e^x = \sum_{n=0}^{\infty} \frac{x^n}{n!}$, $\forall\ x \in \mathbb{R}$. (This will be covered in Chapter 4.)

1.4 Continuous Functions

Definition 1.4.1 *Let $f : A \to \mathbb{R}$, and let $c \in A$. We say that f is* **continuous at** *c if*

$$\lim_{x \to c} f(x) = f(c).$$

Technically, f is **continuous at** *c if for every $\epsilon > 0$ there exists a $\delta > 0$ such that*

$$|f(x) - f(c)| < \epsilon$$

whenever $|x - c| < \delta$ and $x \in A$.

Remark 1.4.1 (1) Equivalently, it can be defined as follows. If x is any point of A satisfying $|x - c| < \delta$, then $|f(x) - f(c)| < \epsilon$.

(2) If f fails to be continuous at c, then we say that f is **discontinuous at** c.

Theorem 1.4.1 (Continuity Test) *If c is a cluster point of A, then three conditions must hold for f to be continuous at c:*

1) f must be defined at c, i.e. $f(c)$ exists.

2) the limit of f at c must exist, i.e. $\lim_{x \to c} f(x)$ exists.

3) these two values must be equal, i.e. $\lim_{x\to c} f(x) = f(c)$.

Theorem 1.4.2 (Sequential Criterion for Continuity) *A function* $f : A \to \mathbb{R}$ *is continuous at the point* $c \in A$ *if and only if, for every sequence* $\{x_n\}$ *in* A *that converges to* c*, the sequence* $\{f(x_n)\}$ *converges to* $f(c)$. *That is,*

$$\lim_{n\to\infty} f(x_n) = f\left(\lim_{n\to\infty} x_n\right).$$

Definition 1.4.2 f *is said to be* **right-continuous** *or* **continuous from the right** *at* c *if* $\lim\limits_{x\to c^+} f(x) = f(c)$. *Similarly,* f *is* **left-continuous** *or* **continuous from the left** *at* c *if* $\lim\limits_{x\to c^-} f(x) = f(c)$.

Definition 1.4.3 *Let* $A \subset \mathbb{R}$ *and let* $B \subset \mathbb{R}$ *be a subset of* A*, we say that* f *is* **continuous on the set** B *if* f *is continuous at every point of* B.

Example 1.4.1 The following functions are continuous on the given set:

(1) The constant function $f(x) = b$ is continuous on $\mathbb{R}$.

(2) $g(x) = x^n$, $n \in \mathbb{N}$, is continuous on $\mathbb{R}$.

(3) $h(x) = 1/x$ is continuous on $\mathbb{R}^+ = \{x \in \mathbb{R} : x > 0\}$.

(4) $h(x) = 1/x$ is not continuous at $x = 0$.

(5) $\psi(x) = \mathrm{sgn}(x)$ is not continuous at 0.

Example 1.4.2 Define $f(x) = x - \lfloor x \rfloor$. Here, $\lfloor x \rfloor$ is the greatest integer less than or equal to x. Therefore, $f(x)$ denotes the part under the decimal point of x. We would like to investigate whether this function is continuous at $x = 1$ or not, using the Continuity Test. Note that

(1) $f(1) = 0$

(2) $\lim_{x\to 1^+} f(x) = 0$, $\lim_{x\to 1^-} f(x) = 1$.

Therefore, $f(x)$ is not continuous at $x = 1$ since the limit at $x = 1$ cannot be defined. In general, $f(x)$ is discontinuous at all integers, but actually right-continuous. Observe that $f(1.01) = 0.01$,$f(0.99) = 0.99$. Since $\lim_{x\to 1} f(x)$ does not exist, $f(x)$ is not continuous at $x = 1$. Refer to Figure 1.5.

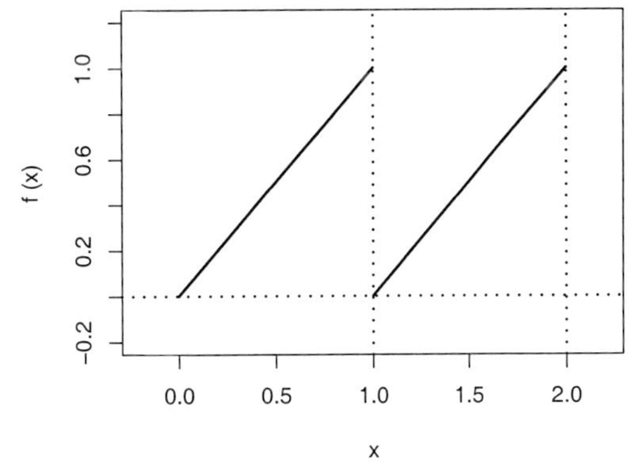

Figure 1.5: Graph of $f(x) = x - \lfloor x \rfloor$

Example 1.4.3 (1) Define

$$f(x) = \begin{cases} \sin\frac{\pi}{x}, & x \neq 0, \\ 0, & x = 0. \end{cases}$$

We will show that it is discontinuous at $x = 0$ by the formal definition of continuity. First, note that the function has the value of $(-1)^k$ at $x = 1/(k+0.5)$ $(k = \pm1, \pm2, \pm3, \cdots)$. Thus, if we choose $\epsilon = 0.5$, we cannot get any δ that satisfies

$$|x| < \delta \text{ implies } |f(x) - f(0)| < \epsilon\,(= 0.5)\ .$$

Since there are too many numbers in the form of $1/(k+0.5)$ in $(-\delta, \delta)$. Therefore, f is not continuous at $x = 0$.

(2) Let $f(x) = x\sin(1/x)$ for $x \neq 0$ (See Figure 1.6). Since f is not defined at $x = 0$, the function f cannot be continuous at $x = 0$. However, we know that $\lim_{x\to 0} x\sin(1/x) = 0$. Thus, if we define $f(x)$ by

$$f(x) = \begin{cases} x\sin\frac{1}{x}, & x \neq 0, \\ 0, & x = 0, \end{cases}$$

then f is continuous at $x = 0$. Alternatively, for a given $\epsilon > 0$, take $\delta = \epsilon/2$. Then we have

$$|x| < \delta \text{ implies } \left| x \sin \frac{1}{x} \right| \leq |x| < \delta < \epsilon.$$

Therefore, f is continuous at $x = 0$.

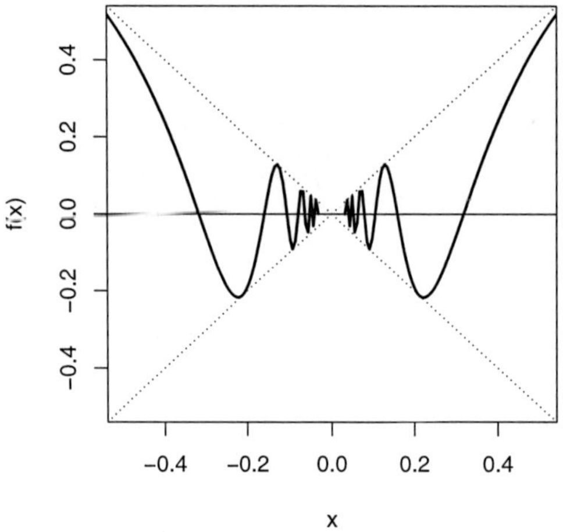

Figure 1.6: Graph of $g(x) = x \sin \frac{1}{x}$

Theorem 1.4.3 *Let $f(x)$ and $g(x)$ be two continuous functions defined on a set $A \subset R$. Then:*

1. *$f(x) \pm g(x)$ and $f(x)g(x)$ are continuous on A.*

2. *$af(x)$ is continuous on A, where a is a constant.*

3. *$f(x)/g(x)$ is continuous on A provided that $g(x) \neq 0$ on A.*

4. *$|f|$ is continuous on A.*

5. *$\sqrt{f}$ is continuous on A.*

Remark 1.4.2 Many functions that we have briefly seen before such as polynomial functions, rational functions, trigonometric functions, exponential and logarithmic functions are continuous. Actually, their graphs are all connected.

Example 1.4.4 Let f and g be functions from A to $\mathbb{R}$ and define the functions $\max(f,g)$ and $\min(f,g)$ from A to $\mathbb{R}$ by

$$\begin{aligned}\max(f,g)(x) &= \max\{f(x),g(x)\},\\ \min(f,g)(x) &= \min\{f(x),g(x)\}.\end{aligned}$$

Then two functions are continuous wherever both f and g are continuous.

Theorem 1.4.4 *Supposes that $f: A \to \mathbb{R}$ is continuous on A, and $g: B \to \mathbb{R}$ is continuous on B. If $f(A) \subset B$, then the composite function $h(= g\circ f): A \to \mathbb{R}$ defined as $h(x) = g(f(x))$ is continuous on A. That is, if f and g are continuous at c, then*

$$\lim_{x\to c} f(g(x)) = f\left(\lim_{x\to c} g(x)\right) = f\left(g\left(\lim_{x\to c} x\right)\right) = f(g(c)).$$

Example 1.4.5 By Theorem 1.4.4, it is easy to see that

1. $\sqrt{|x|}$ is continuous.
2. $\sqrt{x^2+1}$ is continuous.

Example 1.4.6 Define $f:\mathbb{R}\to\mathbb{R}$ and $g:\mathbb{R}\to\mathbb{R}$ like as follows.

$$f(x) = \begin{cases} \frac{1}{x}, & x\neq 0\\ 0, & x=0\end{cases}, \quad g(x) = \begin{cases} \frac{1}{x}, & x\neq 0\\ 0, & x=0\end{cases}$$

Then, $(g\circ f)(x) = x$. Therefore, $g\circ f:\mathbb{R}\to\mathbb{R}$ is continuous at every point. However, f and g are not continuous at $x=0$.

Theorem 1.4.5 *Suppose that $f: A\to\mathbb{R}$ is continuous and strictly monotone increasing (decreasing) on $[a,b] \in A$. Then f^{-1} is continuous and strictly monotone increasing (decreasing) on $f([a,b])$.*

Example 1.4.7 (1) $f(x) = x^3$ is strictly monotone increasing for all $x \in \mathbb{R}$. Thus, it has a unique inverse given by $f^{-1}(y) = y^{1/3}$ that is also continuous.

(2) $f(x) = \sin x$ is strictly monotone increasing for all $x \in [-\pi/2, \pi/2]$. Thus, it has the inverse $f^{-1}(y) = \arcsin y = \sin^{-1} y$.

$$\lim_{x \to \frac{1}{2}} \sin^{-1}(x) = \sin^{-1}\left(\frac{1}{2}\right) = \frac{\pi}{6}.$$

(3) For $x \in [0, \pi]$,

$$\lim_{x \to \frac{\pi}{2}} \cos^{-1}(\cos x) = \cos^{-1}\left(\lim_{x \to \frac{\pi}{2}} \cos x\right) = \cos^{-1}\left(\cos \frac{\pi}{2}\right) = \frac{\pi}{2}.$$

Functions that are continuous on intervals have a number of important properties.

Definition 1.4.4 *A function $f : A \to \mathbb{R}$ is said to be* **bounded** *on A if there exists a constant $M > 0$ such that $|f(x)| \leq M$ for all $x \in A$.*

Theorem 1.4.6 *If f is continuous on the closed bounded interval $[a, b]$, then f is bounded on $[a, b]$.*

Theorem 1.4.7 *(a)* **(Maximum-Minimum Theorem)** *If f is continuous on $[a, b]$, then f has minimum and maximum values on $[a, b]$. That is, there exist points $x_1, x_2 \in [a, b]$ such that $m = f(x_2) \leq f(x) \leq f(x_1) = M$ for all $x \in [a, b]$, where m and M denote minimum and maximum values on $[a, b]$, respectively.*

(b) **(Intermediate Value Theorem)** *If f is continuous on $[a, b]$ and k is between $f(a)$ and $f(b)$ then there exists $c \in (a, b)$ such that $f(c) = k$.*

(c) **(Location of Roots Theorem)** *Let $f : [a, b] \to \mathbb{R}$ be continuous and suppose that $f(a) < 0 < f(b)$. Then there exists a point c in (a, b) such that $f(c) = 0$.*

Example 1.4.8 (1) Let $f(x) = x^2 - x - 1$. Since $f(1) = -1$, $f(2) = 1$ and $f(x)$ is continuous on $[1, 2]$, there exists $c \in [1, 2]$ such that $f(c) = 0$ by the Intermediate Value Theorem or Location of Roots Theorem.

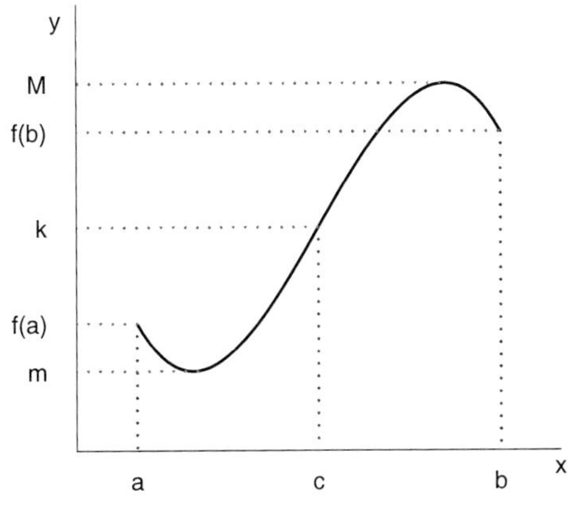

Figure 1.7: $f([a,b]) = [m, M]$ and Intermediate Value Theorem

(2) Using the Intermediate Value Theorem, we can show that every positive number has a positive nth root, where n is an integer ≥ 2. Suppose that $k > 0$ and $n \in \mathbb{N}$. Let $f(x) = x^n$. Then $f(0) = 0 < k$. Furthermore, if $b = k+1$, then

$$b^n = (1+k)^n \geq 1 + kn > k$$

by Bernoulli's inequality. Thus $f(b) > k$. Since f is continuous, we conclude that there exists $c \in (0, b)$ such that $f(c) = k$. Thus $c^n = k$ and c is an nth root of k.

(3) Let $f(x) = x - \cos x$. Then $f(0) = -1$ and $f(\pi/2) = \pi/2$. Since f is continuous, there exists some $x \in (0, \pi/2)$ such that $x = \cos x$ by the Intermediate Value Theorem.

1.5 Suggested Exercises

1. Prove the following inequalities, for $n \geq 1$ using the Binomial Theorem :

   (a) $2^n \geq 1 + n$.

   (b) $2^{1/n} \leq 1 + \frac{1}{n}$.

2. Find the inverse function of

$$f(x) = \frac{1}{1 + 2^{-x}}, \quad -\infty < x < \infty.$$

3. If $f_0(x) = x/(x+1)$ and $f_{n+1} = f_0 \circ f_n$ for $n = 0, 1, 2, \ldots$, find a formula for $f_n(x)$ and use mathematical induction to prove it.

4. Assume that f is an even function, g is an odd function, and both f and g are defined on the entire real line $\mathbb{R}$. Which of the following (where defined) are even? odd?

   (a) $f \cdot g$.

   (b) $f \circ g$.

   (c) $g \circ f$.

   (d) g/f.

5. We would like to find out the limit of the sequence $\{a_n\} = 1/\sqrt{n}$.

   (a) Determine a, the limit of $\{a_n\}$.

   (b) To show that $\lim_{n \to \infty} (1/\sqrt{n}) = a$, given any $\epsilon > 0$ we have to find N such that $n > N$ implies that $|1/\sqrt{n} - a| < \epsilon$. What can we take for N?

6. Graph the function

$$f(x) = \begin{cases} \lfloor x \rfloor, & x \geq 0, \\ \lceil x \rceil, & x < 0. \end{cases}$$

7. Determine the following limits :

   (a) $\lim_{x \to 0} \dfrac{e^x - 1}{\sin x}$.

   (b) $\lim_{x \to 0} e^{\sin x}$.

(c) $\lim_{x\to 0} \frac{\sin x}{|x|}$.

8. Suppose that the range of g lies in the domain of f so that the composite $f \circ g$ is defined. If f and g are one-to-one, can anything be said about $f \circ g$? Give reasons for your answer.

9. Determine the limit of the following sequence : $\{a_n\} = \left\{\left(1 + \frac{(-1)^n}{n}\right)\left(2 - \frac{\cos^3 n}{n^2}\right)\right\}$, i.e.

$$\lim_{n\to\infty} \left(1 + \frac{(-1)^n}{n}\right)\left(2 - \frac{\cos^3 n}{n^2}\right).$$

(You can determine the limit and use basic properties of the limit covered in the class to prove it.)

10. Evaluate $\tan\left(\arcsin \frac{\sqrt{3}}{2}\right)$.

11. If $\lim_{x\to a}[f(x) + g(x)] = 2$ and $\lim_{x\to a}[f(x) - g(x)] = 1$, find $\lim_{x\to a}[f(x)g(x)]$.

12. Let

$$h(x) = \begin{cases} x^2 & x < 2, \\ 3 & x = 2, \\ 2 & x > 2. \end{cases}$$

Show that

(a) $\lim_{x\to 2} h(x) \neq 4$.

(b) $\lim_{x\to 2} h(x) \neq 3$.

(c) $\lim_{x\to 2} h(x) \neq 2$.

13. The formula $\lim_{\theta\to 0} \frac{\sin\theta}{\theta}$ can be generalized. If $\lim_{x\to c} f(x) = 0$ and $f(x)$ is never zero in an open interval containing the point $x = c$, except possible c itself, then

$$\lim_{x\to c} \frac{\sin f(x)}{f(x)} = 1.$$

Evaluate

(a) $\lim_{x\to 0} \frac{\sin x^2}{x}$.

(b) $\lim_{x\to -1} \frac{\sin(x^2 - x - 2)}{x + 1}$.

(c) $\lim_{x\to 0} \dfrac{\sin(1-\cos x)}{x}$.

14. We would like to show that $\lim_{x\to 2} x^2 + x = 6$ using the precise definition of limit. Find a $\delta > 0$ so that $|x-2| < \delta$ implies that

    (a) $|x^2 + x - 6| < 1$.

    (b) $|x^2 + x - 6| < 1/n$ for a given $n \in \mathbb{N}$.

    (c) $|x^2 + x - 6| < \epsilon$ for a given $\epsilon > 0$.

15. Evaluate

$$\lim_{x\to 0} \frac{\sqrt[3]{1+cx}-1}{x},$$

where c is a nonzero constant.

16. Find $\lim_{x\to\infty} e^{-x}\sin x$.

17. Find the following limit

$$\lim_{x\to 0} \frac{x\csc 2x}{\cos 5x}.$$

18. Consider the function $f(x) = \cos(1/x)$ for $x > 0$. Show that $\lim_{x\to 0} f(x)$ does not exist.

19. If the function f is defined by

$$f(x) = \begin{cases} 0 & \text{if } x \text{ is rational,} \\ 1 & \text{if } x \text{ is irrational.} \end{cases}$$

Show that $\lim_{x\to 0} f(x)$ does not exist.

20. Consider the formula

$$f(x) = \lim_{n\to\infty} \frac{x^n}{1+x^n}.$$

Let $D = \{x : f(x) \in \mathbb{R}\}$. Calculate $f(x)$ for all $x \in D$ and determine where $f : D \to \mathbb{R}$ is continuous.

21. Using the Squeeze theorem, we would like to prove that $\lim_{x\to 0^+} x^{3/2} = 0$:

    (a) First, establish the inequality $x < x^{1/2} \le 1$ for $0 < x \le 1$.

    (b) Then, complete your proof.

22. Suppose the function $f : \mathbb{R} \to \mathbb{R}$ has the property that there is some $M > 0$ such that $|f(x)| \leq M|x|^2$ for all x. Prove that

$$\lim_{x\to 0} f(x) = 0, \quad \text{and} \quad \lim_{x\to 0} \frac{f(x)}{x} = 0.$$

23. Show that $\displaystyle\lim_{x\to 0} \frac{e^x - 1}{x} = 1$, and find $\displaystyle\lim_{x\to 0} \frac{a^x - 1}{2x}$, where $a > 0$.

24. Evaluate

    (a) $\displaystyle\lim_{x\to 1/2} \sin^{-1} x$

    (b) $\displaystyle\lim_{x\to \infty} \tan^{-1} x$

25. Define $\sinh x = \dfrac{e^x - e^{-x}}{2}$, $\cosh x = \dfrac{e^x + e^{-x}}{2}$, and $\tanh x = \dfrac{\sinh x}{\cosh x}$. Find $\tanh^{-1} x$.

26. Show that $2^x = 3x$ for some $x \in (0, 1)$.

27. Show that the function $f(x) = 2\log x + \sqrt{x} - 2$ has a solution in the interval $[1, 2]$.

Chapter 2

Differentiation

In this chapter, we begin by presenting the concept of the derivative as an extension of the limits introduced in Chapter 1, and then examine several aspects of differentiation.

2.1 The derivative

Definition 2.1.1 (Tangent Line) *The slope of the curve* $y = f(x)$ *at the point* $P(x_0, f(x_0))$ *is the number*

$$m = \lim_{h \to 0} \frac{f(x_0 + h) - f(x_0)}{h} \quad \textit{(provided the limit exists).}$$

The **tangent line** *to the curve at* P *is the line through* P *with this slope.*

Definition 2.1.2 *The* **derivative** *of a function* f *at a point* x_0, *denoted by* $f'(x_0)$, *is*

$$f'(x_0) = \lim_{h \to 0} \frac{f(x_0 + h) - f(x_0)}{h}$$

provided this limit exists.

Definition 2.1.3 *Let* $I \subset \mathbb{R}$ *be an interval, let* $f : I \to \mathbb{R}$, *and let* $c \in I$. *We say that a real number* L *is the* **derivative of** f **at** c *if given any* $\epsilon > 0$ *there exists* $\delta(\epsilon) > 0$ *such that if* $x \in I$ *satisfies* $0 < |x - c| < \delta(\epsilon)$, *then*

$$\left| \frac{f(x) - f(c)}{x - c} - L \right| < \epsilon.$$

In this case we say that f *is* **differentiable** *at* c, *and we write* $f'(c)$ *for* L.

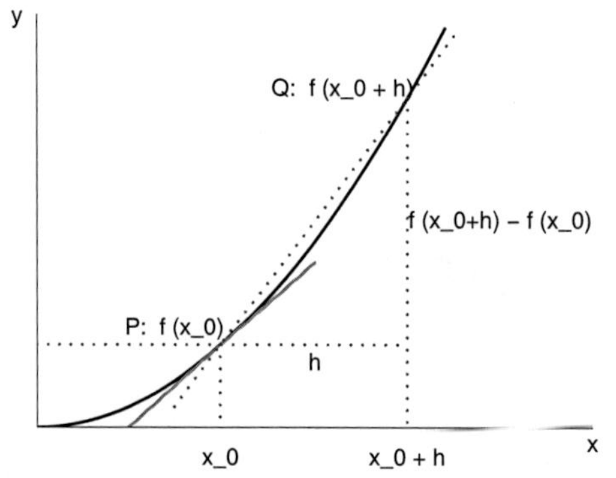

Figure 2.1: Tangent line, i.e. the limit of secant $\overline{PQ}$ when $h \to 0$

Remark 2.1.1 It is also common to use the notation

$$\frac{d}{dx} f(x)\bigg|_{x=c} = f'(c).$$

We thus have

$$f'(c) = \lim_{h \to 0} \frac{f(c+h) - f(c)}{h}.$$

By putting $x = c + h$, above formula can be written as

$$f'(c) = \lim_{x \to c} \frac{f(x) - f(c)}{x - c}.$$

Example 2.1.1 (1) Let $f(x) = x^2$ for each $x \in \mathbb{R}$. Then, for any $c \in \mathbb{R}$, we have $f'(c) = 2c$.

(2) Let $f(x) = \dfrac{1}{x}$. Then for any $c \in \mathbb{R}^+$, we have $f'(c) = -\dfrac{1}{c^2}$.

(3) Let $f(x) = \sqrt{x}$. Then for any $c \in \mathbb{R}^+$, we have $f'(c) = \dfrac{1}{2\sqrt{c}}$.

Remark 2.1.2 (1) The derivative of the function $f(x)$ with respect to the variable x is the function f' defined by

$$f'(x) = \lim_{h\to 0} \frac{f(x+h)-f(x)}{h} \quad \text{provided the limit exists.}$$

$$f'(x) = \frac{d}{dx}f = \frac{dy}{dx} = \lim_{\Delta x\to 0} \frac{\Delta y}{\Delta x}.$$

(2) If f is differentiable at c, then for every sequence $\{x_n\}$ in I that converges to c with $x_n \neq c$ for all n, the sequence

$$\left\{\frac{f(x_n)-f(c)}{x_n - c}\right\}$$

converges to $f'(c)$.

(3) A differential coefficient $f'(c)$ means the slope (gradient) of a tangent line for function f at $x = c$. Look at Figure 2.2. The differential coefficient $f'(1)$ at $P : (1, 1)$ is

$$f'(1) \approx \frac{f(1+h)-f(1)}{(1+h)-1}, \quad \text{for very small } h.$$

In general, the tangent line of function f at $x = c$ is given by

$$y = f(c) + f'(c)(x-c).$$

Theorem 2.1.1 *If $f : I \to \mathbb{R}$ is differentiable at $c \in I$, then f is continuous at c.*

Proof For all $x \in I$, $x \neq c$, we have

$$f(x) - f(c) = \left(\frac{f(x)-f(c)}{x-c}\right)(x-c).$$

Since $f'(c)$ exists, we have

$$\lim_{x\to c}(f(x)-f(c)) = \lim_{x\to c}\left(\frac{f(x)-f(c)}{x-c}\right)\lim_{x\to c}(x-c) = 0,$$

which concludes that f is continuous at c.

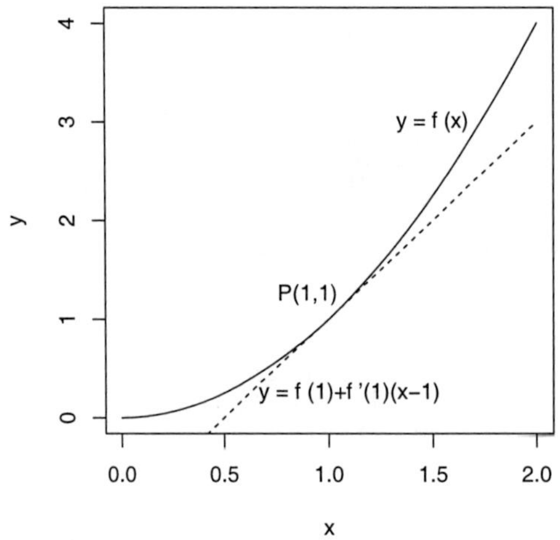

Figure 2.2: $y = f(x)$ and its tangent line at $P(1,1)$.

Remark 2.1.3 The converse is not true as shown in the next example but the contraposition is true. That is, if f is not continuous at $x = c$, then f is not differentiable at $x = c$.

Example 2.1.2 The converse of Theorem 2.1.1 is not true. That is, the continuity of f at a point does not assure the existence of the derivative at that point. For example, if $f(x) = |x|$ for $x \in \mathbb{R}$, then the derivative at 0 does not exist and therefore the function is *not differentiable* at 0. Hence, the continuity at a point c is *not* a sufficient condition for the existence of the derivative at c.

For this function,

$$\lim_{h \to 0+} \frac{1}{h}\{f(h) - f(0)\} = \lim_{h \to 0+} \frac{1}{h}\{+h\} = 1,$$

$$\lim_{h \to 0-} \frac{1}{h}\{f(h) - f(0)\} = \lim_{h \to 0-} \frac{1}{h}\{-h\} = -1.$$

So,

$$\lim_{h \to 0} \frac{1}{h}\{f(h) - f(0)\}$$

does not exist. Therefore, this function is not differentiable at $x = 0$ (but differentiable at other points). Note that the slope of tangent line of $|x|$ for $x > 0$ is 1, whereas the slope for $x < 0$ is -1, as shown in Figure 2.3. Hence $f(x) = |x|$ is not differentiable at $x = 0$ but continuous everywhere.

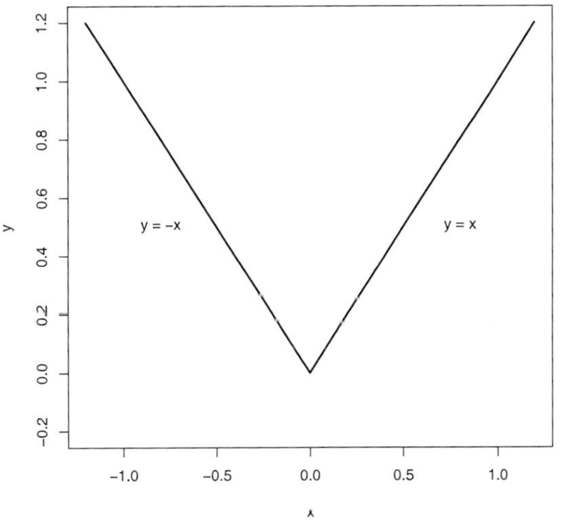

Figure 2.3: Non-differentiability of $f(x) = |x|$ at $x = 0$

Example 2.1.3 Define $f : \mathbb{R} \to \mathbb{R}$ by $f(x) = x\sin(1/x)$ if $x \neq 0$ and $f(0) = 0$. Determine whether or not f is differentiable at $x = 0$.

For $x \neq 0$ we have

$$\frac{f(x) - f(0)}{x - 0} = \frac{x \sin \frac{1}{x}}{x} = \sin \frac{1}{x}.$$

We know that $\lim_{x \to 0} \sin \frac{1}{x}$ does not exist. Thus f is not differentiable at $x = 0$.

Example 2.1.4 Differentiation of a constant function, a linear function and a quadratic function.

(1) $f(x) = k$ (a constant). Then,

$$f'(x) = \lim_{h\to 0} \frac{1}{h}\{f(x+h) - f(x)\} = \lim_{h\to 0} \frac{1}{h}\{k - k\} = 0.$$

(2) $f(x) = x$. Then,

$$f'(x) = \lim_{h\to 0} \frac{1}{h}\{f(x+h) - f(x)\} = \lim_{h\to 0} \frac{1}{h}\{x + h - x\} = 1.$$

(3) $f(x) = x^2$. Then,

$$f'(x) = \lim_{h\to 0} \frac{1}{h}\{f(x+h) - f(x)\} = \lim_{h\to 0} \frac{1}{h}\{(x+h)^2 - x^2\} = \lim_{h\to 0} \frac{1}{h}\{2xh + h^2\} = 2x.$$

Theorem 2.1.2 *Suppose that $f : I \to \mathbb{R}$ and $g : I \to \mathbb{R}$ are differentiable at $c \in I$. Then*

(1) If $k \in \mathbb{R}$, the the function kf is differentiable at c and

$$(kf)'(c) = k \cdot f'(c).$$

(2) The function $f \pm g$ is differentiable at c and

$$(f \pm g)'(c) = f'(c) \pm g'(c).$$

(3) **(Product Rule)** *The function fg is differentiable at c and*

$$(fg)'(c) = f(c)g'(c) + g(c)f'(c).$$

(4) **(Quotient Rule)** *If $g(c) \neq 0$, then the function f/g is differentiable at c and*

$$\left(\frac{f}{g}\right)'(c) = \frac{f'(c)g(c) - f(c)g'(c)}{[g(c)]^2}.$$

Example 2.1.5 To illustrate the use of Theorem 2.1.2, we show that for any $n \in \mathbb{N}$, if $f(x) = x^n$, then $f'(x) = nx^{n-1}$. We prove it by mathematical induction.

When $n = 1$ we have $f(x) = x$, so that

$$f'(x) = \lim_{h\to 0} \frac{x + h - x}{h} = 1,$$

and the formula holds. Now suppose that the formula holds for $n = k$, $k \in \mathbb{N}$. That is, if $f(x) = x^k$, then $f'(x) = kx^{k-1}$. The function $g(x) = x^{k+1} = x^k \cdot x$. By using the product rule of Theorem 2.1.2, we have

$$g'(x) = kx^{k-1} \cdot x + x^k \cdot 1 = (k+1)x^k.$$

Thus the formula holds for $n = k+1$, and by mathematical induction we conclude that it holds for all $n \in \mathbb{N}$.

Alternatively, using the Binomial theorem we can prove it directly :

$$\begin{aligned} f'(x) &= \lim_{h\to 0} \frac{(x+h)^n - x^n}{h} \\ &= \lim_{h\to 0} \frac{\sum_{k=0}^{n} \binom{n}{k} x^k h^{n-k} - x^n}{h} \\ &= \lim_{h\to 0} \frac{\binom{n}{1} x^{n-1} h + \cdots + \binom{n}{n} h^n}{h} \\ &= nx^{n-1}. \end{aligned}$$

Example 2.1.6 Let n be a negative integer and let $f(x) = x^n$ for $x \neq 0$. Note that $-n$ is positive and if $g(x) = x^{-n}$, then $f(x) = 1/g(x)$. Thus, using the quotient rule of Theorem 2.1.2, we have $f'(x) = nx^{n-1}$.

Example 2.1.7 The derivatives of some transcendental functions are

(1) $\dfrac{d}{dx} e^x = e^x$.

Note that

$$\frac{d}{dx} e^x = \lim_{h\to 0} \frac{1}{h}(e^{x+h} - e^x) = e^x \lim_{h\to 0} \frac{1}{h}(e^h - 1).$$

Take $e^h - 1 = u$. Then, we have $e^h = u + 1$, $h = \log(u+1)$. Thus we have

$$\lim_{h\to 0} \frac{e^h - 1}{h} = \lim_{u\to 0} \frac{u}{\log(u+1)} = \lim_{u\to 0} \frac{1}{\frac{1}{u}\log(u+1)} = \lim_{u\to 0} \frac{1}{\log(u+1)^{\frac{1}{u}}} = \lim_{u\to 0} \frac{1}{\log e} = 1.$$

Therefore, $\dfrac{d}{dx} e^x = e^x$.

(2) $\dfrac{d}{dx}\log x = \dfrac{1}{x}$.

Since $\lim_{h\to 0}(1+h)^{\frac{1}{h}} = e(= 2.71828\ldots)$, we have

$$\begin{aligned}
\frac{d}{dx}\log x &= \lim_{h\to 0}\frac{1}{h}\{\log(x+h)-\log x\} = \lim_{h\to 0}\frac{1}{h}\left\{\log\left(1+\frac{h}{x}\right)\right\} \\
&= \frac{1}{x}\lim_{h\to 0}\frac{x}{h}\left\{\log\left(1+\frac{h}{x}\right)\right\} \\
&= \frac{1}{x}\lim_{h\to 0}\left\{\log\left(1+\frac{h}{x}\right)^{\frac{x}{h}}\right\} = \frac{1}{x}\log e = \frac{1}{x}.
\end{aligned}$$

(3) $\dfrac{d}{dx}\sin x = \cos x$.

Since $\lim_{h\to 0}\frac{\sin h}{h} = 1$ and $\sin(a+b) = \sin a\cos b + \cos a \sin b$, we have

$$\begin{aligned}
\frac{d}{dx}\sin x &= \lim_{h\to 0}\frac{1}{h}\{\sin(x+h)-\sin x\} \\
&= \lim_{h\to 0}\frac{1}{h}\{\sin x\cos h + \cos x \sin h - \sin x\} \\
&= \lim_{h\to 0}\frac{1}{h}\sin x(\cos h - 1) + \lim_{h\to 0}\frac{1}{h}\cos x\sin h = \cos x.
\end{aligned}$$

(4) $\dfrac{d}{dx}\cos x = -\sin x$.

Since $\lim_{h\to 0}\frac{\sin h}{h} = 1$ and $\cos(a+b) = \cos a\cos b - \sin a\sin b$, we have

$$\begin{aligned}
\frac{d}{dx}\cos x &= \lim_{h\to 0}\frac{1}{h}\{\cos(x+h)-\cos x\} \\
&= \lim_{h\to 0}\frac{1}{h}\{\cos x\cos h - \sin x\sin h - \cos x\} \\
&= -\lim_{h\to 0}\frac{1}{h}\sin x\sin h = -\sin x.
\end{aligned}$$

(5) $\dfrac{d}{dx}\tan x = \dfrac{d}{dx}\dfrac{\sin x}{\cos x} = \sec^2 x$.

(6) $\dfrac{d}{dx}a^x = a^x\log a, \quad \text{for } a > 0$.

(7) $\dfrac{d}{dx}\log_a x = \dfrac{1}{x \log a}$, for $a > 0$.

Example 2.1.8 If $f_1, f_2, \ldots, f_n$ are functions on an interval I to $\mathbb{R}$ that are differentiable at $c \in I$, then

(1) $\left(\sum_{i=1}^n f_i\right)'(c) = \sum_{i=1}^n f_i'(c)$.

(2) $(f_1 f_2 \cdots f_n)'(c) = f_1'(c) f_2(c) \cdots f_n(c) + f_1(c) f_2'(c) \cdots f_n(c) + \cdots + f_1(c) f_2(c) \cdots f_n'(c)$.

(3) $(f^n)'(c) = n(f(c))^{n-1} f'(c)$.

Remark 2.1.4 (Leibniz Notation) We sometimes write Df for f'. Then we have

$$D(f+g) = Df + Dg, \quad D(fg) = (Df)\cdot g + f\cdot(Dg).$$

Or, we denote

$$\frac{d}{dx}(f(x)g(x)) = \frac{df}{dx}(x)g(x) + f(x)\frac{dg}{dx}(x),$$

which is due to Leibniz.

Theorem 2.1.3 (Chain Rule) *Let I and J be intervals in $\mathbb{R}$, let $f : I \to \mathbb{R}$ and $g : J \to \mathbb{R}$, where $f(I) \subset J$, and let $c \in I$. If f is differentiable at c and g is differentiable at $f(c)$, then the composite function $g \circ f$ is differentiable at c and*

$$(g \circ f)'(c) = g'(f(c)) \cdot f'(c).$$

In Leibniz notation, if $y = g(u)$ and $u = f(x)$, then

$$\frac{dy}{dx} = \frac{dy}{du}\cdot\frac{du}{dx},$$

where dy/du is evaluated at $u = f(x)$.

Proof We write

$$\frac{(g\circ f)(x) - (g\circ f)(c)}{x-c} = \frac{g(f(x)) - g(f(c))}{f(x)-f(c)} \cdot \frac{f(x)-f(c)}{x-c}.$$

By talking the limit of both sides at $x \to c$, we obtain the result.

Alternatively, note that $(g\circ f)'(x) = \lim_{h\to 0}\frac{1}{h}\{g(f(x+h)) - g(f(x))\}$. If $k = f(x+h) - f(x)$, $y = f(x)$, then

$$\begin{aligned}(g\circ f)'(x) &= \lim_{h\to 0}\frac{1}{h}\{g(y+k) - g(y)\}\\ &= \lim_{h\to 0}\frac{1}{k}\{g(y+k) - g(y)\}\frac{k}{h}.\end{aligned}$$

Since $\lim_{h\to 0}\{f(x+h) - f(x)\} = 0$,

$$\begin{aligned}(g\circ f)'(x) &= \lim_{h\to 0,\ k\to 0}\frac{1}{k}\{g(y+k) - g(y)\}\frac{1}{h}\{f(x+h) - f(x)\}\\ &= g'(y)f'(x) = g'(f(x))f'(x).\end{aligned}$$

Remark 2.1.5 (1) The only problem in the proof of Chain Rule is that $f(x) - f(c)$ may be zero even when $x - c \neq 0$, or $k = f(x+h) - f(x) = 0$. Thus we define a new function $h : J \to \mathbb{R}$ by

$$h(y) = \begin{cases} \frac{g(y)-g(f(c))}{y-f(c)}, & \text{if } y \neq f(c),\\ g'(f(c)), & \text{if } y = f(c),\end{cases}$$

and see that h is continuous at $f(c)$. Thus, $\lim_{x\to c}(h\circ f)(x) = h(f(c)) = g'(f(c))$. Therefore

$$\begin{aligned}(g\circ f)'(c) &= \lim_{x\to c}\frac{(g\circ f)(x) - (g\circ f)(c)}{x-c}\\ &= \lim_{x\to c}[h\circ f(x)]\cdot\lim_{x\to c}\frac{f(x) - f(c)}{x-c}\\ &= g'(f(c))\cdot f'(c).\end{aligned}$$

(2) Note that

$$\frac{\Delta y}{\Delta x} = \frac{\Delta y}{\Delta u}\cdot\frac{\Delta u}{\Delta x},$$

where $\Delta y = g(u+\Delta u) - g(u)$ and $\Delta u = f(x+\Delta x) - f(x)$. Since $\Delta u \to 0$ as $\Delta x \to 0$, we have

$$\begin{aligned}\frac{dy}{dx} &= \lim_{\Delta x\to 0}\frac{\Delta y}{\Delta x}\\ &= \lim_{\Delta x\to 0}\frac{\Delta y}{\Delta u}\cdot\frac{\Delta u}{\Delta x}\\ &= \lim_{\Delta x\to 0}\frac{\Delta y}{\Delta u}\cdot\lim_{\Delta x\to 0}\frac{\Delta u}{\Delta x}\\ &= \lim_{\Delta u\to 0}\frac{\Delta y}{\Delta u}\cdot\lim_{\Delta x\to 0}\frac{\Delta u}{\Delta x}\\ &= \frac{dy}{du}\frac{du}{dx},\end{aligned}$$

which is written in **Leibniz** notation.

(3) Let $h(x) = (g \cdot f)(x) = g(f(x))$ and let $u = f(x)$. Then by the Chain Rule, we have

$$\frac{d}{dx}h(x) = \frac{d}{dx}g(u) = \frac{d}{du}g(u)\frac{du}{dx} = g'(u)f'(x) = g'(f(x))f'(x).$$

(4) We can extend the Chain Rule to a composite of three (or more) functions; for example

$$(f \circ g \circ h)'(x) = f'[g(h(x))]g'(h(x))h'(x).$$

In Leibniz notation, if we put $v = h(x)$, $u = g(v)$ and $y = f(u) = f[g(h(x)))]$, we have

$$\frac{dy}{dx} = \frac{dy}{du} \times \frac{du}{dv} \times \frac{dv}{dx}.$$

Example 2.1.9 Find the derivative of $h(x) = 2(3x^2 + 2x + 1)^4$.

If $f(x) = 3x^2 + 2x + 1(= y), g(y) = 2y^4$, then

$$h(x) = (g \circ f)(x), \quad f^{'}(x) = 6x + 2, \quad g^{'}(y) = 8y^3.$$

$$\therefore \ h^{'}(x) = 8(3x^2 + 2x + 1)^3(6x + 2).$$

Example 2.1.10 Consider $f(x) = x\sin(1/x)$ if $x \neq 0$ and $f(0)$. Using the fact that the derivative of $\sin x$ is $\cos x$ for all $x \in \mathbb{R}$, we compute the derivative of f at any point $x \neq 0$. Let $h(x) = \sin(1/x)$, so that $f(x) = xh(x)$. Then for $x \neq 0$ the chain rule gives us

$$h'(x) = \left[\cos\frac{1}{x}\right]\left[-x^{-2}\right].$$

Thus we have

$$f'(x) = xh'(x) + h(x) = -\frac{1}{x}\cos\frac{1}{x} + \sin\frac{1}{x}, \quad \forall\ x \neq 0.$$

Since $f'(0)$ does not exist as in the previous example, f is a continuous function that has derivatives at every real x except $x = 0$.

Example 2.1.11 (1) $(f^n)'(x) = n(f(x))^{n-1}f'(x)$, $n \in \mathbb{N}^+$.

(2) $(\sqrt{1 - x^2})' = -\dfrac{x}{\sqrt{1 - x^2}}$, $x \neq \pm 1$.

(3) Suppose that f is defined by

$$f(x) = \begin{cases} x^2 \sin(1/x), & \text{for } x \neq 0, \\ 0, & \text{for } x = 0. \end{cases}$$

Then, $f'(x) = 2x\sin(1/x) - \cos(1/x)$ for $x \neq 0$, and $f'(0) = 0$ by applying the definition of derivative. Therefore, the derivative f' of f exists at all $x \in \mathbb{R}$. However, the function f' is discontinuous at $x = 0$. Thus, a function f that is differentiable at every point of $\mathbb{R}$ needs not have a continuous derivative f'.

Theorem 2.1.4 (Inverse Function Rule) *Let $f : I \to J$, where I is an interval and J is the image of I, i.e. $J = f(I)$, be a function such that*

1. *f is strictly monotone on I,*
2. *f is differentiable on I,*
3. *$f'(x) \neq 0$ on I.*

Then f^{-1} is differentiable at $b = f(a)$, $a \in I$, and

$$(f^{-1})'(b) = \frac{1}{f'(a)} = \frac{1}{f'(f^{-1}(b))},$$

which is shown in Figure 2.4.

Remark 2.1.6 Suppose that the inverse function of function $f : I \to \mathbb{R}$ is $g : J \to \mathbb{R}$, $J = f(I)$. If f is differentiable on I and $f'(x) \neq 0$ for $x \in I$, then g is differentiable on J and

$$g'(x) = \frac{1}{f'(f^{-1}(x))}.$$

Example 2.1.12 (1) The function $f : \mathbb{R} \to \mathbb{R}$ defined by $f(x) = x^5 + 4x + 3$ is continuous and strictly monotone increasing (since it is the sum of two strictly increasing functions). In addition, $f(1) = 8$ and $f'(x) = 5x^4 + 4$ is never zero. Thus, we obtain $(f^{-1})'(8) = 1/f'(f^{-1}(8)) = 1/9$.

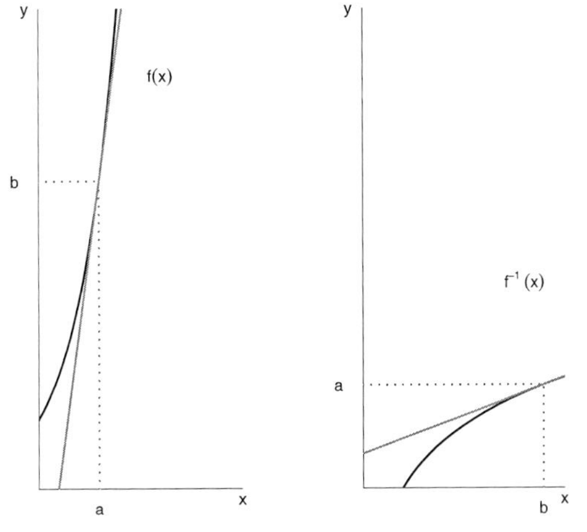

Figure 2.4: The graphs of inverse functions and their tangent lines

(2) Let $f(x) = x^n$ for $x \in [0, \infty)$ and let $n \in \mathbb{N}$ be even. Then, its inverse function $g(x) = f^{-1}(x) = x^{1/n}$. Then,

$$(f^{-1})'(x) = \frac{1}{f'(f^{-1}(x))} = \frac{1}{n\,(f^{-1}(x))^{n-1}} = \frac{1}{n}x^{1/n-1}.$$

When $n \in \mathbb{N}$ is odd, $f(x) = x^n$, $x \in \mathbb{R}$, we also get $(f^{-1})'(x) = \frac{1}{n}x^{1/n-1}$.

(3) Let $r = m/n$ be a positive rational number, and let $h(x) = x^r$ for $x \in [0, \infty)$. Then $h(x)$ is the composition of the function $f(x) = x^m$ and $g(x) = x^{1/n}$. That is, $h(x) = f(g(x))$. Thus,

$$h'(x) = f'(g(x))g'(x) = m(x^{1/n})^{m-1}\frac{1}{n}x^{(1/n)-1} = \frac{m}{n}x^{(m/n)-1} = rx^{r-1}.$$

(4) Let $f(x) = \sin x$, $-\frac{\pi}{2} < x < \frac{\pi}{2}$. Find $(\sin^{-1})'(0.5)$. Note that

$$\sin(\pi/6) = 0.5, \quad \sin^{-1}(y) = 1/\sin'(x) = 1/\cos x.$$

$$\therefore \ (\sin^{-1})'(0.5) = 1/\cos(\pi/6) = 2/\sqrt{3}.$$

In addition, by writing $\sin^{-1}(x) = a$.

$$(f^{-1}(x))' = \frac{1}{f'(f^{-1}(x))} = \frac{1}{\cos(\sin^{-1}(x))} = \frac{1}{\cos a} = \frac{1}{\sqrt{1-x^2}}$$

(5) If $f(x) = e^x$, then $f^{-1}(x) = \log x$. Thus, the derivative of $\log x$ is obtained by

$$(\log x)' = (f^{-1}(x))' = \frac{1}{f'(f^{-1}(x))} = \frac{1}{e^{\log x}} = \frac{1}{x}.$$

Example 2.1.13 We can also generalize the derivative rule of $f(x) = x^a, a \in \mathbb{R}$ in the following way:

(1) Consider $f(x) = x^{\frac{m}{n}}$. Then by denoting $y = x^{\frac{m}{n}}$, i.e. $y^n = x^m$ and differentiating both sides, we have

$$\begin{aligned} \frac{d}{dx}y^n &= \frac{d}{dx}x^m = mx^{m-1}, \\ \frac{d}{dy}y^n \times \frac{dy}{dx} &= mx^{m-1}, \quad \text{(by Chain Rule)} \\ \frac{dy}{dx}ny^{n-1} &= mx^{m-1}. \end{aligned}$$

Therefore, we obtain

$$\frac{dy}{dx} = \frac{m}{n} \times \frac{x^{m-1}}{y^{n-1}} = \frac{m}{n} \times x^{m-1} \times x^{-\frac{m}{n}(n-1)} = \frac{m}{n}x^{\frac{m}{n}-1}.$$

(2) $\frac{d}{dx}x^a = ax^{a-1}$, $x > 0$, for any real number a. We can prove this easily by using the fact that $x^a = e^{a\log_e x}$ for $x > 0$ and the Chain Rule.

Example 2.1.14 (Logarithmic Differentiation)

(1) $(\log f)' = \frac{f'}{f}$ or $f' = f \cdot (\log f)'$

(2) Consider $f(x) = a^x$ and find the derivative of $f(x)$, i.e. a^x.

$$\log f(x) = x \log a \Rightarrow \frac{d}{dx}\log f = \frac{d}{dx}(x \log a) = \log a.$$

Thus,

$$\frac{d}{dx}\log f = \log a \Rightarrow \frac{f'}{f} = \log a \Rightarrow f' = f \log a = a^x \log a.$$

(3) Consider $f(x) = x^x$ and find the derivative of $f(x)$, i.e. $\frac{d}{dx}(x^x)$.

$$\begin{aligned} f &= x^x. \\ \log f &= x\log x. \\ (\log f)' &= \log x + x\frac{1}{x} = \log x + 1. \\ \frac{f'}{f} &= \log x + 1. \\ f' &= (\log x + 1)x^x. \end{aligned}$$

Alternatively, we write $f = e^{x\log x}$ and differentiate f using Chain Rule.

Differentiation of the Parametric Equation

Definition 2.1.4 *If x and y are given as functions*

$$x = f(t), \ y = g(t)$$

over an interval of t, then the set of points $(x, y) = (f(t), g(t))$ defined by these equations is called a **parametric curve** . *The equations are* **parametric equations** *for the curve.*

Theorem 2.1.5 *(1) If $x = f(t)$ and $y = g(t)$ are differentiable at t and $f'(t) \neq 0$, then*

$$\frac{dy}{dx} = \frac{dy/dt}{dx/dt} = \frac{g'(t)}{f'(t)}.$$

(2) If the equation $x = f(t)$ and $y = g(t)$ define y as a twice-differentiable function of x, then at any point where $dx/dt \neq 0$ and $y' = dy/dx$,

$$\frac{d^2y}{dx^2} = \frac{dy'/dt}{dx/dt}.$$

Proof (1)

$$\frac{dy}{dx} = \lim_{h\to 0}\frac{g(t+h)-g(t)}{f(t+h)-f(t)} = \lim_{h\to 0}\frac{(g(t+h)-g(t))/h}{(f(t+h)-f(t))/h} = \frac{g'(t)}{f'(t)}.$$

(2)

$$\frac{d^2y}{dx^2} = \frac{d}{dx}(y') = \frac{dy'/dt}{dx/dt}.$$

Example 2.1.15 Let $x = 3\cos t$, $y = 2\sin t$ $(0 \le t \le 2\pi)$. For $t = \pi/3, (x, y) = (3/2, \sqrt{3})$. Then, the slope of the tangent line at $t = \pi/3$ is

$$\left.\frac{dy}{dx}\right|_{t=\pi/3} = \frac{g'(t)}{f'(t)} = -\frac{2\cos t}{3\sin t} = -\frac{2}{3\sqrt{3}}.$$

Implicit Differentiation

For some cases we need to find $\dfrac{dy}{dx}$ for the implicit function of x and y expressed in the form of $G(x,y)=0$, we can find $\dfrac{dy}{dx}$ by differentiating both sides of the equation with respect to x.

Implicit Differentiation :

(1) Differentiate both sides of the equation with respect to x, treating y as a differentiable function of x.

(2) Collect the terms with $\dfrac{dy}{dx}$ on one side of the equation.

(3) Solve for $\dfrac{dy}{dx}$.

Example 2.1.16 (1) The parametric equation of Example 2.1.15 can be expressed in the following parametric equation.

$$\frac{x^2}{9}+\frac{y^2}{4}=1.$$

Then, the slope of the tangent line at the point $(3/2,\sqrt{3})$ can be found by differentiating both sides of equation with respect to x :

$$\begin{aligned}
\frac{d}{dx}\frac{x^2}{9}+\frac{d}{dx}\frac{y^2}{4} &= 0.\\
\therefore\ \frac{d}{dx}\frac{x^2}{9}+\frac{d(y^2/4)}{dy}\frac{dy}{dx} &= 0.\ \text{(by Chain Rule)},\\
\frac{2}{9}x+\frac{2}{4}y\frac{dy}{dx} &= 0.\\
\therefore\ \frac{dy}{dx}=-\frac{4}{9}\frac{x}{y} &= -\frac{2}{3\sqrt{3}}.
\end{aligned}$$

(2) Equation of a circle with a radius of $1: x^2+y^2=1$ which we can write as $y^2=1-x^2$. So $y=\pm\sqrt{1-x^2}$. If we consider the positive case, $y=\sqrt{1-x^2}$, its derivative is given by

$$\frac{dy}{dx}=\frac{-x}{\sqrt{1-x^2}}=-\frac{x}{y}.$$

If we use implicit differentiation and apply chain rule, we have

$$\begin{aligned} \frac{d}{dx}x^2 + \frac{d}{dx}y^2 &= 0. \\ 2x + \frac{d}{dx}y^2 &= 0. \\ \Rightarrow\ 2x + \frac{d}{dy}y^2 \times \frac{dy}{dx} &= 0. \text{ (by Chain Rule)} \\ \Rightarrow\ 2x + 2y\frac{dy}{dx} &= 0. \\ \therefore\ \frac{dy}{dx} &= \frac{-x}{y}, \end{aligned}$$

which yields the same answer. On the other hand, we can consider a parametric equation for the unit circle, $x = \sin\theta$ and $y = \cos\theta$. Then, we also get

$$\frac{dy}{dx} = \frac{dy/d\theta}{dx/d\theta} = \frac{-\sin\theta}{\cos\theta} = -\frac{x}{y}.$$

(3) Consider $y = a^x$. We obtain $\dfrac{dy}{dx} = a^x \log a$ by taking the logarithm of $y = a^x$ and using implicit differentiation.

(4) $y^4 + xy^2 - 2 = 0$. In this case, it is not easy to solve for y as a function of x. Instead, we use implicit differentiation to find $\dfrac{dy}{dx}$.

$$\begin{aligned} \frac{d}{dx}y^4 + \frac{d}{dx}xy^2 - \frac{d}{dx}2 &= 0 \\ \frac{d}{dy}y^4\frac{dy}{dx} + y^2 + x\frac{d}{dy}y^2\frac{dy}{dx} &= 0 \\ 4y^3\frac{dy}{dx} + y^2 + 2xy\frac{dy}{dx} &= 0 \\ \therefore\ \frac{dy}{dx} &= -\frac{y^2}{4y^3 + 2xy}. \end{aligned}$$

(5) $y^3 + xy^2 + 1 = 0$. Find $\dfrac{dy}{dx}$.

Remark 2.1.7 In (4) of Example 2.1.16, it can be also written as

$$4y^3dy + y^2dx + 2xydy = 0.$$

2.2 Application of Differentiation

Linear and Quadratic Approximations

If f is differentiable, then $f'(x) = \lim_{h \to 0} \frac{1}{h}\{f(x+h) - f(x)\}$, implying that

$$f'(x) \approx \frac{1}{h}\{f(x+h) - f(x)\}, \quad \text{for small } h.$$

Therefore, we can think of the following approximation of $f(x)$,

$$f(x+h) \approx f(x) + f'(x)h, \quad \text{for small } h.$$

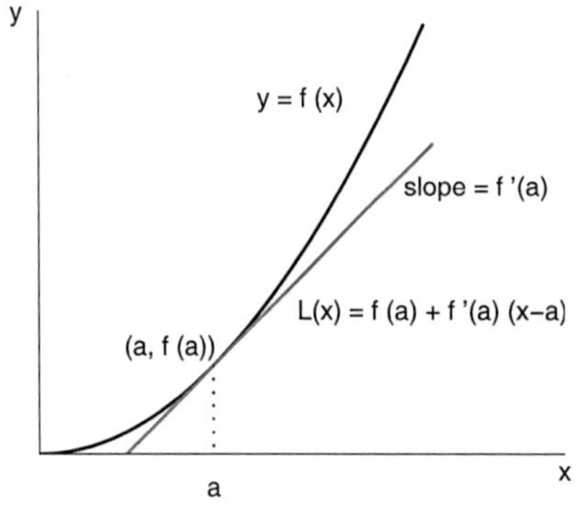

Figure 2.5: The tangent to the curve as a linear approximation to the curve at $x = a$.

The tangent line at $x = x_0$, $L(x) = f(x_0) + f'(x_0)(x - x_0)$ approximates $f(x)$ in that $L(x_0) = f(x_0)$ and $L'(x_0) = f'(x_0)$. It provides a good approximation to $f(x)$ near x_0. As we move away from x_0, however, the approximation becomes less accurate. This linear approximation is also known as the **linearization**, actually **a first order Taylor series expansion** of $f(x)$ at $x = a$.

$$f(x) \approx L(x) = f(x_0) + f'(x_0)(x - x_0).$$

Example 2.2.1 Let $f(x) = \log x$, and let $x_0 = 1$. Then, since

$$f(1) = \log 1 = 0\ ;\quad f'(1) = \left.\frac{1}{x}\right|_{x=1} = 1,$$

$\log x$ is approximated by $f(1) + f'(1)(x-1)$ at $x_0 = 1$, i.e. $\log x \approx x - 1$. In addition, we have

$$\log(1+u) \approx u,$$

at $u_0 = 0$ by taking $x - 1 = u$ and $x_0 - 1 = u_0 = 0$.

Example 2.2.2 Consider a linear approximation to the following functions at $x_0 = 0$:

1. $f(x) = e^x \approx 1 + x$.
2. $f(x) = \sin x \approx x$.
3. $f(x) = \cos x \approx 1$.
4. $f(x) = \sqrt{x+1} \approx 1 + \frac{x}{2}$.
5. $f(x) = (1+x)^r \approx 1 + rx$.

Example 2.2.3 (1) Find the value of $\sqrt{1.1}$ using the linear approximation of $f(x) = \sqrt{x}$. If $f(x) = \sqrt{x}$, then $f'(x) = \frac{1}{2\sqrt{x}}$. Therefore,

$$\sqrt{x+h} \approx \sqrt{x} + \frac{1}{2\sqrt{x}}(x + h - x).$$

By taking $x = 1$ and $h = 0.1$,

$$\sqrt{1.1} \approx 1 + \frac{1}{2}0.1 = 1.05.$$

By calculator, we obtain $\sqrt{1.1} = 1.0488$.

(2) Find the value of $\log 0.9$ using the linear approximation of $f(x) = \log x$. If $f(x) = \log x$, then $f'(x) = \frac{1}{x}$. Therefore,

$$\log(x+h) \approx \log x + \frac{h}{x}.$$

By taking $x = 1$ and $h = -0.1$,

$$\log 0.9 \approx \log 1 + \frac{-0.1}{1} = -0.1.$$

By calculator, $\log 0.9 = -0.1054$.

Remark 2.2.1 Using the linear approximation, we can find out the limit of functions. For example,

$$\lim_{x\to 0}\frac{e^x-1}{x}=\lim_{x\to 0}\frac{1+x-1}{x}=1,$$

$$\lim_{x\to 0}\frac{\sin x}{x}=\lim_{x\to 0}\frac{x}{x}=1.$$

The quadratic approximation is more complicated, so it is used when higher accuracy is needed.

Definition 2.2.1 (Quadratic Approximation) *Suppose that a function $f(x)$ is twice-differentiable at $x=a$. Then, the quadratic equation*

$$y=f(a)+f'(a)(x-a)+f''(a)(x-a)^2/2$$

has the intersection at $(a, f(a))$ with f, also the first derivative and second derivative at $x=a$ are equal to $f'(a)$ and $f''(a)$ respectively. Therefore, we have

$$f(x)\approx f(a)+f'(a)(x-a)+f''(a)(x-a)^2/2$$

around $x=a$.

Consider $f(x)=ax^2+bx+c$. Then, $f'(x)=2ax+b$, and $f''(x)=2a$. At $x=0$, $f(0)=c$, $f'(0)=b$ and $f''(0)=2a$. Therefore,

$$f(x)=f(0)+f'(0)x+\frac{f''(0)}{2}x^2.$$

Example 2.2.4 Find the linearization and the quadratic approximation of $f(x)=1+xe^{-x}$ at $x=0$.

Note that $f(0)=1,\ f'(0)=1,\ f''(0)=-2$. Thus, the linear approximation at $x=0$ is $y=1+x$, and the quadratic approximation at $x=0$ is $y=1+x-x^2$.

Example 2.2.5 Consider a quadratic approximation to the following functions at $x_0=0$, i.e. $x\approx 0$:

1. $f(x)=e^x\approx 1+x+\frac{1}{2}x^2$.
2. $f(x)=\sin x\approx x$.
3. $f(x)=\cos x\approx 1-\frac{x^2}{2}$.
4. $f(x)=\log(1+x)\approx x-\frac{1}{2}x^2$.
5. $f(x)=(1+x)^r\approx 1+rx+\frac{r(r-1)}{2}x^2$.

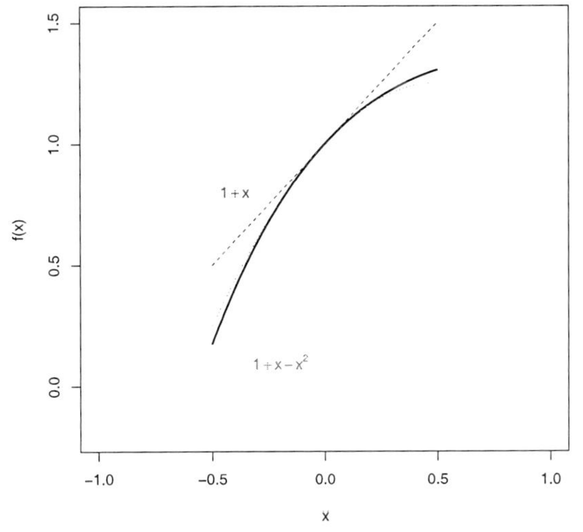

Figure 2.6: The graph of $f(x) = 1+xe^{-x}$ and its linearization and quadratic approximation at $x = 0$.

Mean Value Theorem

Definition 2.2.2 *Let f be a function with domain D. Then f has an* **global (absolute) maximum** *value on D at a point c, if*

$$f(x) \le f(c), \quad \forall x \in D,$$

and an **global (absolute) minimum** *value on D at c, if*

$$f(x) \ge f(c), \quad \forall x \in D.$$

Definition 2.2.3 *Let f be defined on an interval I, and let $c \in I$. Then*

- *f has a* **local maximum** *$f(c)$ at $c \in I$ if*

  $$f(x) \le f(c) \text{ for all } x \text{ in some open interval containing } c.$$

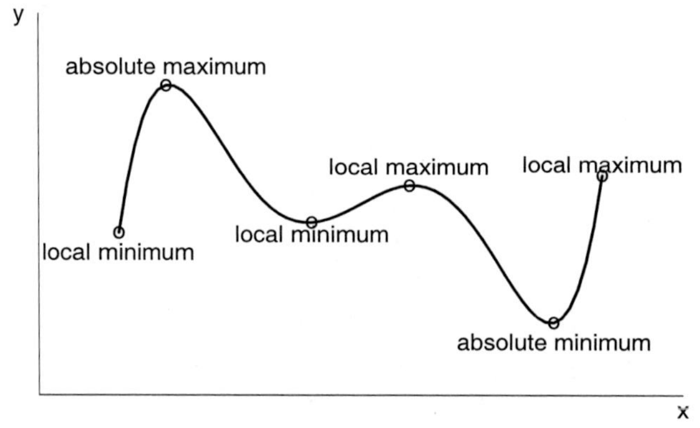

Figure 2.7: Maxima and Minima

- *f has a* **local minimum** *$f(c)$ at $c \in I$ if*

$$f(x) \geq f(c) \quad \textit{for all } x \textit{ in some open interval containing } c.$$

- *f has a* **local extremum** *at c if $f(c)$ is either a local maximum or a local minimum.*

Theorem 2.2.1 (Local Extremum Theorem) *Let f be defined on an interval $[a,b]$. If f has a local extremum at c, where $a < c < b$, and f is differentiable at c, then*

$$f'(c) = 0.$$

Proof If f has a local maximum at c, $f(c) \geq f(x)$ for all x near c. Thus we have

$$f'(c) = \lim_{h \to 0+} \frac{1}{h}\{f(c+h) - f(c)\} \leq 0, \tag{2.2.1}$$

$$f'(c) = \lim_{h \to 0-} \frac{1}{h}\{f(c+h) - f(c)\} \geq 0. \tag{2.2.2}$$

Since f is differentiable at c, the left and right derivative at c exists and are must be equal. Hence, together with (2.2.1) and (2.2.2), we must have $f'(c) = 0$. The proof for a local minimum is similar.

Remark 2.2.2 (1) The Local Extremum Theorem applies only if the function is differentiable at a local extremum. For example, the function

$$f(x) = |x|, \quad x \in [-1, 1],$$

has a local minimum 0 at 0, but f is not differentiable at 0.

(2) The converse is not true. That is, a point where the derivative vanishes is not necessarily a local extremum. For example, the function

$$f(x) = x^3, \quad x \in [-1, 1],$$

does not have a local extremum at 0, although $f'(0) = 0$.

(3) To determine the maximum and the minimum of a function f which is continuous on $[a, b]$, examine

(a) the points x in (a, b) where $f'(x) = 0$.

(b) the endpoints a and b.

(c) the points where f is not differentiable. And,

(d) compare all these values. The least is the minimum, and the greatest is the maximum.

(4) The Local Extremum Theorem does not make any assertion about a local extremum that occurs at a point c that is one of the end-points of the interval $[a, b]$.

Example 2.2.6 An immediate consequence of the Local Extremum Theorem is the following criterion for finding all the extrema of 'well-behaved functions' on a closed interval.

- Let f be continuous on the closed interval $[a, b]$ and differentiable on the open interval (a, b). Then the extrema of f on $[a, b]$ can occur only at a, at b, or at points c in (a, b) where $f'(c) = 0$.

Using this criterion, we can easily find the extrema of the function $f(x) = x^3 - 12x - 5$, $x \in [-3, 3]$.

Theorem 2.2.2 (Rolle's Theorem) *Suppose that f is continuous on the closed interval $[a, b]$ and differentiable on (a, b). If $f(a) = f(b)$, then there exists at least one point c with $a < c < b$ such that $f'(c) = 0$.*

Remark 2.2.3 Rolle's Theorem is one of the most important results in mathematical analysis.

(1) In geometric interpretation, Rolle's Theorem means that if the line joining the points $(a, f(a))$ and $(b, f(b))$ on the graph of f is horizontal, then so is the tangent to the graph at some point c in (a, b), as in Figure 2.8.

(2) There may be more than one point c in (a, b) at which $f'(c) = 0$.

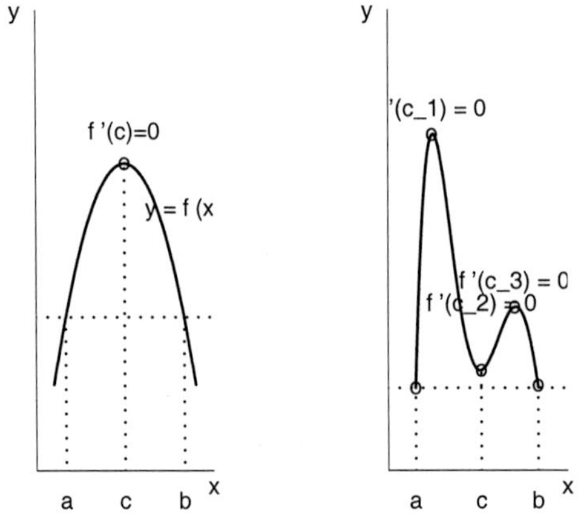

Figure 2.8: Rolle's Theorem

Proof If f is a constant function at $[a, b]$, then $f'(c) = 0$ for every $c \in (a, b)$. If f is not a constant function on $[a, b]$, f has either the maximum or minimum (or both) of f on $[a, b]$ which is different from $f(a) = f(b)$. Since one of the extremum occurs at some point c with $a < c < b$, the Local Extremum Theorem applied to the point c implies that $f'(c) = 0$.

We now generalize Rolle's Theorem and assert that there is always a point where the tangent line to the graph is parallel to the straight line joining the end-points, known as the **Mean Value Theorem**, so-called since

$$\frac{f(b)-f(a)}{b-a}$$

can be regarded as the mean value of the derivatives between a and b.

Theorem 2.2.3 (Mean Value Theorem) *Let f be continuous on $[a,b]$ and differentiable on (a,b). Then there exists at least one point c in (a,b) such that*

$$f'(c)=\frac{f(b)-f(a)}{b-a}.$$

Remark 2.2.4 The idea of the proof is as follows. We define $h(x)$ to be the vertical distance from the chord (i.e., the straight line that connects $(a,f(a))$ and $(b,f(b))$) to the curve $f(x)$. Then $h(a)$ and $h(b)$ are both zero, which satisfies the condition of Rolle's Theorem. Applying Rolle's Theorem to h, we obtain the desired result. Refer to Figure 2.9.

Proof The slope of the chord joining the points $(a,f(a))$ and $(b,f(b))$ is

$$m=\frac{f(b)-f(a)}{b-a},$$

and so the equation of the chord is

$$y=m(x-a)+f(a).$$

Let a function $h(x)$ be the vertical height between points with ordinate x on the graph and those on the chord, given by

$$h(x)=f(x)-f(a)-\frac{f(b)-f(a)}{b-a}(x-a).$$

Then $h(a)=h(b)=0$, and h is continuous on $[a,b]$ and differentiable on (a,b). Thus h satisfies all the conditions of Rolle's Theorem. Therefore there exists some point c in (a,b) at which $h'(c)=0$ by Rolle's Theorem. In addition, since $h'(c)=f'(c)-m$, it follows that

$$\begin{aligned} h'(c) &= f'(c)-\frac{f(b)-f(a)}{b-a}=0, \\ f'(c) &= \frac{f(b)-f(a)}{b-a}. \end{aligned}$$

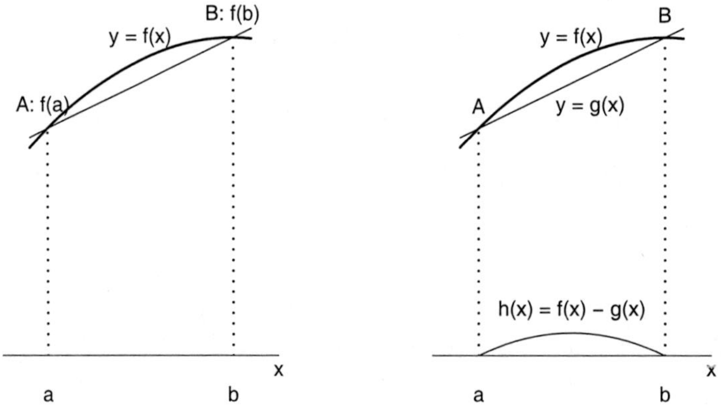

Figure 2.9: The Mean Value Theorem

Example 2.2.7 One important use of the Mean Value is to obtain certain inequalities.

(a) $e^x \geq 1 + x$ for $x \in \mathbb{R}$.

If $x > 0$, we apply the Mean Value Theorem to the function on the interval $[0, x]$. Then for some c with $0 < c < x$ we have

$$e^x - e^0 = e^c(x - 0) > x.$$

A similar argument establishes the same strict inequality for $x < 0$. If $x = 0$, we have the equality with both sides equal to 1.

(b) $-x \leq \sin x \leq x$ for all $x \geq 0$.

If we apply the Mean Value Theorem to $\sin x$ on $[0, x]$, $x > 0$, we obtain

$$\sin x - \sin 0 = (\cos c) \cdot (x - 0),$$

for some c between 0 and x. Since $\sin 0 = 0$ and $-1 \leq \cos c \leq 1$, we have $-x \leq \sin x \leq x$.

(c) (Bernoulli's inequality) If $\alpha > 1$, then

$$(1+x)^\alpha \geq 1 + \alpha x \quad \text{for all } x > -1.$$

If $f(x) = (1+x)^\alpha$ then $f'(x) = \alpha(1+x)^{\alpha-1}$ for all $x > -1$. If $x > 0$, we apply the Mean Value Theorem to f on $[0, x]$. Thus, there exists c with $0 < c < x$ such that

$$(1+x)^\alpha - 1 = \alpha(1+c)^{\alpha-1}x > \alpha x.$$

If $-1 < x < 0$, a similar use of the Mean Value Theorem on $[x, 0]$ leads to the same inequality.

Curve Sketching

Theorem 2.2.4 (Zero-derivative Theorem) *Let f be continuous on $[a, b]$ and differentiable on (a, b). If $f'(x) = 0$ for all $x \in (a, b)$, then f is constant on $[a, b]$.*

Remark 2.2.5 Let f and g be continuous on $[a, b]$ and differentiable on (a, b). Suppose that $f'(x) = g'(x)$ for all $x \in (a, b)$. Then there exists a constant C such that $f = g + C$ on $[a, b]$, which is a direct application of Theorem 2.2.4.

Example 2.2.8 We can prove the claim that $f(x) = \lambda e^x$, λ an arbitrary constant, is the only function f that satisfies the differential equation $f'(x) = f(x)$ on $\mathbb{R}$.

For, if $f'(x) = f(x)$, then

$$\begin{aligned} \frac{d}{dx}(e^{-x}f(x)) &= e^{-x}f'(x) - e^{-x}f(x) \\ &= e^{-x}(f'(x) - f(x)) = 0. \end{aligned}$$

Thus, $e^{-x}f(x)$ is just some constant λ. Hence, $f(x) = \lambda e^x$.

Recall that a function $f : I \to \mathbb{R}$ is said to be **increasing** on the interval I if $f(x_1) < f(x_2)$ whenever $x_1 < x_2$, and f is said to be **decreasing** on I if the function $-f$ is increasing on I, that is $f(x_1) > f(x_2)$ whenever $x_1 < x_2$.

Theorem 2.2.5 (Increasing-Decreasing Functions) *Let f be differentiable on an interval I. Then*

(1) *if* $f'(x) \geq 0$ *for all* $x \in I$, *then* f *is increasing on* I.

(2) *if* $f'(x) \leq 0$ *for all* $x \in I$, *then* f *is decreasing on* I.

Proof Since $f(x_2) = f(x_1) + f'(c)(x_2 - x_1)$, for some c between x_1 and x_2 for $x_1 < x_2$ by Mean Value Theorem, $f'(c) > 0$ and $x_2 - x_1 > 0$ implies that

$$f(x_2) > f(x_1).$$

The proof of (2) is similar.

Remark 2.2.6 If the inequalities in Theorem 2.2.5 are replaced by strict inequalities, then the conclusions are also replaced by **strictly** increasing and **strictly** decreasing.

Example 2.2.9 We can show the following inequality using Theorem 2.2.5.

$$\log(1+x) \geq x - \frac{x^2}{2}, \quad x \geq 0.$$

Let $f(x) = \log(1+x) - \left(x - \frac{x^2}{2}\right)$ for $x \geq 0$. Then,

$$f'(x) = \frac{1}{1+x} - (1-x) = \frac{x^2}{1+x} \geq 0, \ \forall\, x \geq 0.$$

Therefore f is increasing on $\mathbb{R}^+ \cup \{0\}$, i.e. $f(x) \geq f(0)$, for all $x \geq 0$.

Theorem 2.2.6 (First Derivative Test for Extrema) *Let* f *be continuous on* $[a, b]$ *and* $c \in (a, b)$. *Assume that* f *is differentiable on* (a, c) *and* (c, b). *Then*

(1) *if there is a neighborhood* $(c-\delta, c+\delta) \subset I$ *such that* $f'(x) \geq 0$ *for* $c-\delta < x < c$ *and* $f'(x) \leq 0$ *for* $c < x < c+\delta$, *then* f *has a local maximum at* c.

(2) *if there is a neighborhood* $(c-\delta, c+\delta) \subset I$ *such that* $f'(x) \leq 0$ *for* $c-\delta < x < c$ *and* $f'(x) \geq 0$ *for* $c < x < c+\delta$, *then* f *has a local minimum at* c.

Remark 2.2.7 If a function f which is continuous at $[a, b]$ and differentiable on (a, b) has a local maximum(minimum) at $x = c$ $(a < c < b)$, then $f'(c) = 0$.

Remark 2.2.8 An interior point of the domain of a function f where f' is zero or undefined is a ***critical point*** of f.

Remark 2.2.9 Steps for curve-sketching (See Figure 2.10) :

1. Check the discontinuities of f.
2. Plot the critical points.
3. Decide the sign of $f'(x)$ in between the critical points.
4. Find the zeros.
5. Determine the behavior at the endpoints.

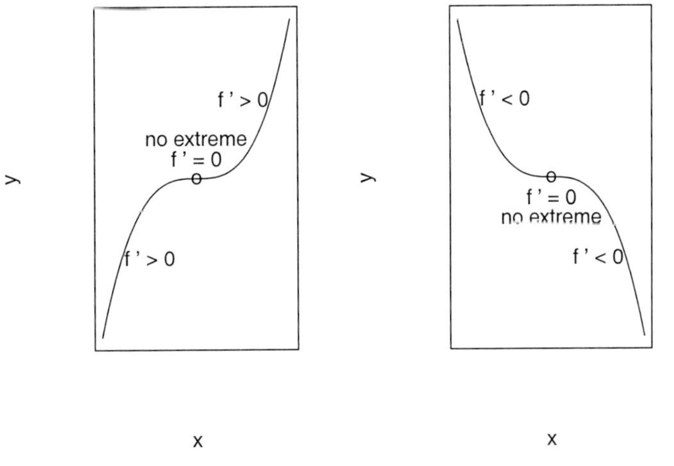

Figure 2.10: The first derivative determines where the curve increases or decreases.

Example 2.2.10 Consider $f(x) = x^3 - 12x - 5$.

(1) No discontinuities.

(2) $f'(x) = 3x^2 - 12$ so, $f'(x) = 0$ at $x = \pm 2$.

(3) At $x = -2$, $f(x) = 11$ that is a local maximum and at $x = 2$, $f(x) = -22$ which is a local minimum. When $|x| > 2$, $f'(x) > 0$, and $f'(x) < 0$ when $|x| < 2$, which implies that $f(x)$ is decreasing between $x = -2$ and $x = 2$, and is increasing when $x > 2$ and $x < -2$.

(4) Find the zeros.

(5) As $x \to \infty$, $f(x) \to \infty$. Similarly, as $x \to -\infty$, $f(x) \to -\infty$.

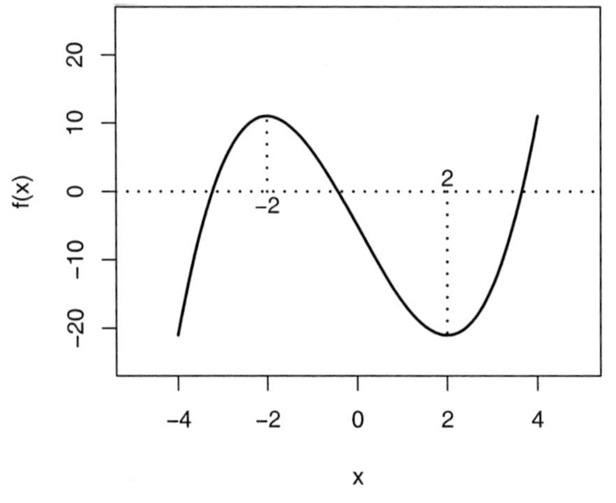

Figure 2.11: Sketch of the function $f(x) = x^3 - 12x - 5$

Convex and Concave Functions

Definition 2.2.4 *Let $I \subset \mathbb{R}$ be an interval. A function $f : I \to \mathbb{R}$ is said to be* **convex** *on I if, for any t satisfying $0 \le t \le 1$ and any points x_1, x_2 in I,*

$$f((1-t)x_1 + tx_2) \le (1-t)f(x_1) + tf(x_2).$$

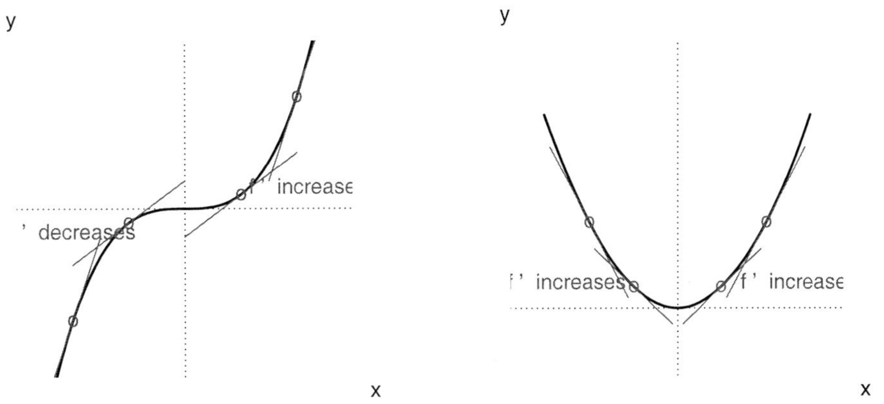

Figure 2.12: Convex vs. Concave

Remark 2.2.10 f is convex (*or* concave up) if the derivative increases as x increases, while f is concave down the slope of the tangent line decreases as x increases, as shown in Figure 2.12.

Theorem 2.2.7 *Let I be an open interval and let $f : I \to \mathbb{R}$ have a second derivative on I. Then f is a convex function on I if and only if $f''(x) > 0$ for all $x \in I$.*

Remark 2.2.11 The sign of the second derivative tells us about convexity or concavity of the graph. That is, f is convex if f' is increasing on (a, b), and f is concave if f' is decreasing on (a, b). When $f'' > 0$, f' is increasing, and f' is decreasing when $f'' < 0$. Thus, the second derivative tells us

(1) whether a critical point is a local maximum or a local minimum, known as the second derivative test.

$f'(x_0)$	$f''(x_0)$	Critical point is a
0	negative	local maximum
0	positive	local minimum

(2) the shape of graph, convexity or concavity.

Definition 2.2.5 *The point where $f''=0$ are called* ***inflection points****. Usually, at these points the graph changes from concave up to down or in the other way.*

Theorem 2.2.8 (Second Derivative Test for Extrema) *Let f be defined on a neighborhood of c, and $f'(c)=0$.*

(a) If $f''(c)>0$, then $f(c)$ is a local minimum of f.

(b) If $f''(c)<0$, then $f(c)$ is a local maximum of f.

Example 2.2.11 Revisit $f(x)=x^3-12x-5$. Since $f''(x)=6x$, f has a point of inflection at $x=0$. See Figure 2.13.

x		-2		0		2	
$f'(x)$	$+$	0	$-$	$-$	$-$	0	$+$
$f''(x)$	$-$	$-$	$-$	0	$+$	$+$	$+$
$f(x)$	increases concave		decreases concave	inflected	decreases convex		increases convex

Example 2.2.12 Consider the standard normal probability density function,

$$\phi(x)=\frac{1}{\sqrt{2\pi}}\exp\left(-\frac{x^2}{2}\right), \quad -\infty<x<\infty.$$

We would like to draw the graph of $\phi(x)$. Note that

$$\begin{aligned}\phi'(x) &= -\frac{1}{\sqrt{2\pi}}xe^{-\frac{x^2}{2}} = -x\,\phi(x),\\ \phi''(x) &= \frac{1}{\sqrt{2\pi}}(x^2-1)e^{-\frac{x^2}{2}}.\end{aligned}$$

Thus,

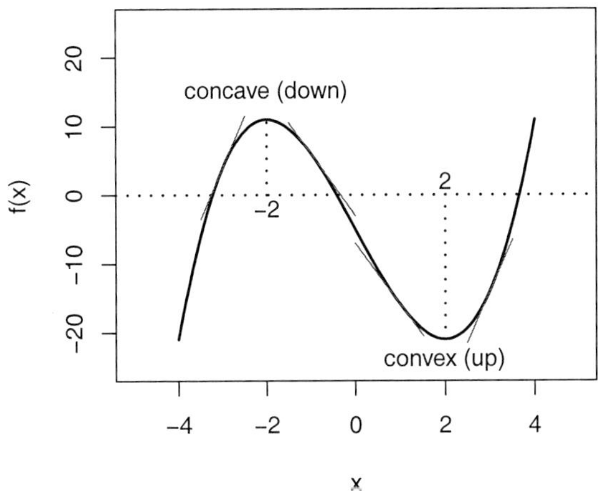

Figure 2.13: Inflection point : $f(x) = x^3 - 12x - 5$, $f''(x) = 0$ at $x = 0$.

x		-1		0		1	
$\phi'(x)$	$+$	$+$	$+$	0	$-$	$-$	$-$
$\phi''(x)$	$+$	0	$-$	$-$	$-$	0	$+$
$\phi(x)$	increases convex	inflection point	increases concave	maximum point	decreases concave	inflection point	decreases convex

(1) f is convex for $x < -1$ and $x > 1$,

(2) f has inflection points at $x = -1$ and $x = 1$,

(3) f is concave for $-1 < x < 1$.

(4) Horizontal asymptote : $y = 0$.

$$\lim_{x \to \infty} \frac{1}{\sqrt{2\pi}} \exp\left(-\frac{x^2}{2}\right) = \lim_{x \to -\infty} \frac{1}{\sqrt{2\pi}} \exp\left(-\frac{x^2}{2}\right) = 0.$$

Sometimes, the normal distribution is referred to as a **bell-shaped** distribution, as shown in Figure 2.14.

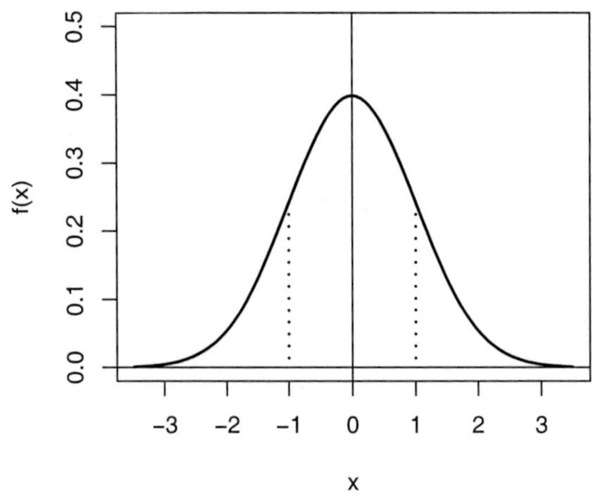

Figure 2.14: Sketch of the standard normal probability density function

L'Hospital's Rule

There are many situations that we would like to evaluate the limit of a quotient of functions,

$$\lim_{x\to c}\frac{f(x)}{g(x)}.$$

Suppose that $\lim_{x\to c} f(x) = L$ and $\lim_{x\to c} g(x) = M$. If $g(x) \neq 0$ for x close to c and $M \neq 0$, then the limit of the quotient f/g is equal to L/M. If both L and M are zero, we can sometimes evaluate the limit of f/g by canceling a common factor in the quotient. In general, when $L = M = 0$, the limit of the quotient f/g is called an **indeterminate form**.

Theorem 2.2.9 (Cauchy's Mean Value Theorem) *Let f and g be continuous on $[a, b]$ and differentiable (a, b) and assume that $g'(x) \neq 0$ for all x in (a, b). Then there exists c in (a, b) such that*

$$f'(c)\cdot\{g(b)-g(a)\} = g'(c)\cdot\{f(b)-f(a)\}.$$

In particular, if $g(b) \neq g(a)$ and $g'(c) \neq 0$, it can be written in the form

$$\frac{f(b) - f(a)}{g(b) - g(a)} = \frac{f'(c)}{g'(c)}.$$

Remark 2.2.12 Note that the Mean Value Theorem is simply the special case of Theorem 2.2.9 in which $g(x) = x$. However, Theorem 2.2.9 is not a simple consequence of the Mean Value Theorem, but the proof of Theorem 2.2.9 is similar to that of the Mean Value Theorem.

Proof Consider the function

$$h(x) = f(x) \cdot \{g(b) - g(a)\} - g(x) \cdot \{f(b) - f(a)\}, \text{ for } x \in [a, b].$$

Then there exists a point c in (a, b) for which $h'(c) = 0$ by Rolle's theorem.

Theorem 2.2.10 (L'Hospital's Rule) *Let $-\infty < a < b < +\infty$, and let $f, g : (a, b) \to \mathbb{R}$ be differentiable functions on (a, b), with $g'(x) \neq 0\ \forall x \in (a, b)$. Suppose that $\lim_{x\to c} f(x) = 0 = \lim_{x\to c} g(x)$. Then,*

$$\lim_{x\to c} \frac{f(x)}{g(x)} \text{ exists and equals } \lim_{x\to c} \frac{f'(x)}{g'(x)},$$

provided that the latter limit exists.

Example 2.2.13 We can evaluate $\displaystyle\lim_{x\to\frac{\pi}{2}} \frac{\cos 3x}{\sin x - e^{\cos x}}$. Let $f(x) = \cos 3x$ and $g(x) = \sin x - e^{\cos x}$, for $x \in \mathbb{R}$. Then f and g are differentiable on $\mathbb{R}$ and $f\left(\frac{\pi}{2}\right) = g\left(\frac{\pi}{2}\right) = 0$. Thus by L'Hospital's rule, we have

$$\lim_{x\to\frac{\pi}{2}} \frac{\cos 3x}{\sin x - e^{\cos x}} = \lim_{x\to\frac{\pi}{2}} \frac{f'(x)}{g'(x)} = \lim_{x\to\frac{\pi}{2}} \frac{-3\sin 3x}{\cos x + \sin x \cdot e^{\cos x}} = 3.$$

Theorem 2.2.11 (L'Hospital's Rule) *Let f and g be differentiable on (b, ∞). Suppose that $\lim_{x\to\infty} f(x) = \lim_{x\to\infty} g(x) = \infty$, and that $g'(x) \neq 0$ for $x \in (b, \infty)$. If*

$$\lim_{x\to\infty} \frac{f'(x)}{g'(x)} = L, \quad \text{where } L \in \mathbb{R},$$

then

$$\lim_{x\to\infty} \frac{f(x)}{g(x)} = L.$$

Example 2.2.14 By L'Hospital's rule, we have

(1) $\lim_{x\to 0+} \frac{\sin x}{\sqrt{x}} = \lim_{x\to 0+} \frac{\cos x}{1/(2\sqrt{x})} = 0.$

(2) $\lim_{x\to 0} \frac{1-\cos x}{x^2} = \lim_{x\to 0} \frac{\sin x}{2x} = \frac{1}{2}.$

(3) $\lim_{x\to 0} \frac{e^x - 1}{x} = 1, \quad \lim_{x\to\infty} \frac{\log x}{x} = 0.$

(4) $\lim_{x\to\infty} e^{-x}x^2 = 0.$

(5) $\lim_{x\to 0+} \frac{\log(\sin x)}{\log x} = \lim_{x\to 0+} \frac{\cos x/\sin x}{1/x} = 1.$

(6) $\lim_{x\to\infty} (1+1/x)^x = \lim_{x\to 0+} (1+x)^{\frac{1}{x}} = e.$

2.3 Suggested Exercises

1. Consider the formula
$$f(x) = \lim_{n\to\infty} \frac{x^n}{1+x^n}.$$
Let $D = \{x : f(x) \in \mathbb{R}\}$. Calculate $f(x)$ for all $x \in D$ and determine where $f : D \to \mathbb{R}$ is continuous.

2. Evaluate $\lim_{x\to 1} \arcsin\left(\frac{1-\sqrt{x}}{1-x}\right)$.

3. Show that
$$\lim_{x\to 0} \frac{e^x - 1}{x} = 1,$$
and find $\lim_{x\to 0} \frac{a^x - 1}{2x}$, where $a > 0$.

4. Let f be a function satisfying $|f(x)| \le x^2$ for $-1 \le x \le 1$. Show that f is differentiable at $x = 0$ and find $f'(0)$. Then, consider the following function
$$f(x) = \begin{cases} x^2 \sin \frac{1}{x}, & x \neq 0, \\ 0, & x = 0. \end{cases}$$
Show that f is continuous at $x = 0$ and that f is differentiable at $x = 0$, and find $f'(0)$.

5. Compute $\lim_{x\to\infty}(\sqrt{x^2+1} - x)$.

6. Suppose f is a function that satisfies the equation
$$f(x+y) = f(x) + f(y) + x^2y + xy^2$$
for all real numbers x and y. Suppose also that
$$\lim_{x\to 0} \frac{f(x)}{x} = 1.$$
(a) Find $f(0)$. (b) Find $f'(0)$. (c) Find $f'(x)$.

7. Find
$$\lim_{h\to 0}(1+2h)^{\frac{1}{h}}, \qquad \lim_{n\to\infty}\left(1 - \frac{1}{2n}\right)^n.$$

8. Show that $2^x = 3x$ for some $x \in (0, 1)$.

9. Find the derivative of the following function

$$f(x) = \frac{x^2 + 3e^x}{2e^x - x}.$$

10. Suppose u and v are differentiable functions of x and that

$$u(1) = 2,\ u'(1) = 0,\ v(1) = 5,\ v'(1) = -1.$$

Find the values of the following derivatives at $x = 1$.

(a) $\frac{d}{dx}\left(uv + \frac{v}{u}\right)$.

(b) $\frac{d}{dx}\left(7v - 2u + \frac{u^2}{v}\right)$.

11. Find dy/dx and d^2y/dx^2 of the following function

$$y = \log(\log x).$$

12. Let $f(x) = x^3 + x$. Find $(f^{-1})'(10)$.

13. Find dy/dx of $xy = \cot(xy)$ by using implicit differentiation.

14. Find dy/dx of following function

$$y = \frac{\cos x}{x} + \frac{x}{\cos x}.$$

15. Find the derivative of y with respect to t, where

$$y = t \log_3 \left(e^{(\sin t) \log 3}\right).$$

16. Show that if $y = A \sin(\log x) + B \cos(\log x)$, where A and B are constants, then

$$x^2 y'' + xy' + y = 0.$$

17. Find the derivative of following function

$$h(x) = x \tan(2\sqrt{x}) + 7.$$

18. Find the derivative of y with respect to θ, where

$$y = \log\left(\frac{\sin\theta\cos\theta}{1+2\log\theta}\right).$$

19. Suppose f is differentiable function such that $f(g(x)) = x$ and $f'(x) = 1 + [f(x)]^2$. Show that $g'(x) = 1/(1+x^2)$. In addition, consider $g(x) = \arctan(x)$, i.e. $\tan^{-1}(x)$, the inverse function of $\tan x$. Provide an estimate of $\arctan(0.1)$ using the linear approximation of $\arctan x$ at $x = 0$.

20. Find y' if $x^y = y^x$, $x, y > 0$.

21. Let f be differentiable on $\mathbb{R}$. Suppose that $f(0) = 0$ and that $1 \le f'(x) \le 2$ for all $x \ge 0$. Prove that $x \le f(x) \le 2x$ for all $x \ge 0$.

22. Let $f(x) = xe^{-\frac{x^2}{2}}$ for $-\infty < x < \infty$. Find the local minimum and the local maximum of $f(x)$.

23. Suppose that a function f satisfies the following conditions for all real values of x and y :

    (a) $f(x+y) = f(x) \cdot f(y)$

    (b) $f(x) = 1 + xg(x)$, where $\lim_{x\to 0} g(x) = 1$.

    Show that the derivative $f'(x)$ exists at every value of x and that $f'(x) = f(x)$. In addition, find the linear approximation of $f(x)$ at $x = 0$.

24. Find y'' if $x^4 + y^4 = 16$.

25. Prove that

    (a) if f is a differentiable even function, then f' is an odd function.

    (b) if f is a differentiable odd function, then f' is an even function.

26. Use the Mean Value Theorem to prove the identity

$$|\sin a - \sin b| \le |a - b| \quad \text{for all } a \text{ and } b.$$

27. Suppose a and c are real numbers, $c > 0$, and f is defined on $[-1, 1]$ by

$$f(x) = \begin{cases} x^a \sin(x^{-c}), & \text{if } x \neq 0, \\ 0, & \text{if } x = 0. \end{cases}$$

Show that

(a) $f(x)$ is continuous if $a > 0$.

(b) $f'(0)$ exists if $a > 1$.

(c) $f'(x)$ is continuous if $a > 1 + c$.

(d) $f''(0)$ exist if $a > 2c$.

28. Let $f(x)$ be a twice-differentiable function that satisfies the following conditions :

$$-f(x) + x^2[f(x)]^3 = 0, \quad \text{and } f(1) = 1.$$

Let $g(x)$ be a quadratic function, $g(x) = m_2(x-1)^2 + m_1(x-1) + m_0$, where m_0, m_1, and m_2 are constants. Suppose that the following conditions are imposed on

$$E(x) = f(x) - g(x).$$

(i) $E(1) = 0$,

(ii) $\lim_{x \to 1} \dfrac{E(x)}{x - 1} = 0$,

(iii) $\lim_{h \to 0} \dfrac{E(h+1) + E(1-h) - 2E(1)}{h^2} = 0$.

Determine constants, m_0, m_1, and m_2.

29. Assume that $f(x)$ is continuous on $[-1, 1]$ and differentiable on (a, b). Also assume that $f(a)$ and $f(b)$ have oppositive signs and that $f'(x) \neq 0$ between a and b. Show that there exists a solution to the equation $f(x) = 0$, and that the solution is unique between a and b.

Chapter 3

Integration

In this chapter we present the basic theory of Riemann integration. Subsequent to Riemann's work, more general approaches to integration were developed such as Stieltjes integration and Lebesgue integration, but we will not cover them in this book. The Riemann integral is commonly used in calculus course and is defined as the limit of the Riemann sums which we will see in detail.

3.1 The Riemann Integral

Definition 3.1.1 (Antiderivative) *A function F is an* **antiderivative** *of f on an interval I if $F'(x) = f(x)$ for all x in I.*

Example 3.1.1 For $\log x$, $x \log x - x$ is an antiderivative since its derivative is $\log x$.

Definition 3.1.2 (Indefinite integral) *If F is an antiderivative of f on an interval I, then the general antiderivative of f on I is $F(x) + C$, where C is an arbitrary constant. The set of all antiderivatives of f is the* **indefinite integral of** *f with respect to x, denoted by*

$$\int f(x)dx = F(x) + C$$

Remark 3.1.1 The symbol $\int$ is **an integral sign**. The function f is the **integrand** of the integral, and x is the **variable** of integration.

Consider the area between a graph and the x-axis for $x \in [a, b]$. The area is called **the integral of** f over $[a, b]$, and denoted by

$$\int_a^b f \quad \text{or} \quad \int_a^b f(x)dx.$$

Then, the area under a curve S can be obtained as follows (See Figure 3.1) :

(1) Divide the region between $[a, b]$ into rectangles,

(2) Add up area of rectangles,

(3) Take limit as rectangles become thin.

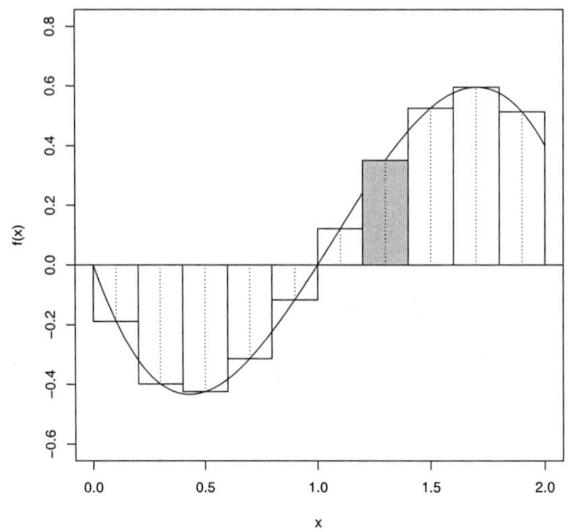

Figure 3.1: Area under a curve; Sum of areas under rectangles

Riemann sum

Definition 3.1.3 *A* **partition** $\boldsymbol{P}$ *of an interval* $[a, b]$ *is a family of a finite number of subintervals of* $[a, b]$*:*

$$\boldsymbol{P} = \{[x_0, x_1], [x_1, x_2], \ldots, [x_{i-1}, x_i], \ldots, [x_{n-1}, x_n]\},$$

where

$$a = x_0 < x_1 < x_2 < \cdots < x_{i-1} < x_i < \cdots < x_{n-1} < x_n = b.$$

The points x_i $(i = 0, \ldots, n)$ *are called the* **partition points** *in* $\boldsymbol{P}$.

Definition 3.1.4 (Riemann sum) *Let* f *be a bounded function on* $[a, b]$, *and* $\boldsymbol{P}$ *a partition of* $[a, b]$ *given by* $P = \{[x_{i-1}, x_i] : 1 \le i \le n\}$. *We denote* m_i *and* M_i *for the quantities*

$$m_i = \min\{f(x) : x \in [x_{i-1}, x_i]\} \quad and \quad M_i = \max\{f(x) : x \in [x_{i-1}, x_i]\}.$$

The the corresponding **lower** *and* **upper Riemann sums** *for* f *on* $[a, b]$ *are*

$$L(f, \boldsymbol{P}) = \sum_{i=1}^{n} m_i(x_i - x_{i-1}) \quad and \quad U(f, \boldsymbol{P}) = \sum_{i=1}^{n} M_i(x_i - x_{i-1}).$$

Remark 3.1.2 In the definition of Riemann sum, min and max need to be replaced by inf (infimum) and sup (supremum) for correctness.

Theorem 3.1.1 (Riemann's Criterion for integrability) *Let* f *be a bounded function on an interval* $[a, b]$.

(1) Then f *is* **integrable** *on* $[a, b]$ *if and only if for each positive number* ϵ, *there is a partition* $\boldsymbol{P}$ *of* $[a, b]$ *for which*

$$U(f, \boldsymbol{P}) - L(f, \boldsymbol{P}) < \epsilon,$$

(2) If f *is integrable on* $[a, b]$, *there is a sequence* $\{\boldsymbol{P}_n\}$ *of partitions of* $[a, b]$ *such that*

$$\lim_{n\to\infty} L(f, \boldsymbol{P}_n) = \lim_{n\to\infty} U(f, \boldsymbol{P}_n) = \int_a^b f,$$

which is called the **Riemann integral of** f.

(3) If there is a sequence $\{\boldsymbol{P}_n\}$ *of partitions of* $[a, b]$ *such that*

$$\lim_{n\to\infty} L(f, \boldsymbol{P}_n) \text{ and } \lim_{n\to\infty} U(f, \boldsymbol{P}_n) \text{ both exist and are equal,}$$

then f *is integrable on* $[a, b]$ *and the common value of these two limits is* $\displaystyle\int_a^b f$.

(4) *Let f be a bounded function on an interval $[a, b]$ and $\{\boldsymbol{P}_n\}$ any sequence of partitions of $[a, b]$ with $\|\boldsymbol{P}_n\| \to 0$, where $\|\boldsymbol{P}_n\| = \max\{\Delta x_i \equiv x_i - x_{i-1},\ i = 1, \ldots, n\}$.*

(a) *If f is integrable on $[a, b]$, then*

$$\lim_{n\to\infty} L(f, \boldsymbol{P}_n) = \lim_{n\to\infty} U(f, \boldsymbol{P}_n) = \int_a^b f.$$

(b) *If there is a number I such that $\lim_{n\to\infty} L(f, \boldsymbol{P}_n)$ and $\lim_{n\to\infty} U(f, \boldsymbol{P}_n)$ both exist and equal to I, then f is integrable on $[a, b]$ and $I = \displaystyle\int_a^b f$.*

Remark 3.1.3 To get the Riemann integral $\displaystyle\int_a^b f(x)dx$, also called a **definite integral**, as in Figure 3.2,

(1) We divide the interval $[a, b]$ into n equal subintervals and denote

$$\boldsymbol{P}_n = \{[x_0, x_1], [x_1, x_2], \ldots, [x_{i-1}, x_i], \ldots, [x_{n-1}, x_n]\},$$

where $x_i = a + \dfrac{(b-a)i}{n}$, $i = 0, 1, \ldots, n$.

(2) We choose m_i to be the minimum and M_i to be the maximum of $f(x)$ on the interval $[x_{i-1}, x_i]$. Since $m_i \le f(x) \le M_i$, for $x \in [x_{i-1}, x_i]$. Let S_i be an area between $x = x_{i-1}$, $x = x_i$, $y = 0$ and $y = f(x)$. Then, we have $m_i \cdot (b-a)/n \le S_i \le M_i \cdot (b-a)/n$.

(3) Therefore, $S = \sum_{i=1}^n S_i$ is between L_n and U_n,

$$L_n = \textstyle\sum_{i=1}^n m_i \cdot (b-a)/n,\ U_n = \sum_{i=1}^n M_i \cdot (b-a)/n.$$

(4) If $\lim_{n\to\infty} U_n = \lim_{n\to\infty} L_n$, then we define the definite integral,

$$S = \int_a^b f(x)dx = \lim_{n\to\infty} U_n = \lim_{n\to\infty} L_n.$$

Example 3.1.2 We may evaluate $\int_a^b f(x)dx$ from the definition of Riemann integral in several cases.

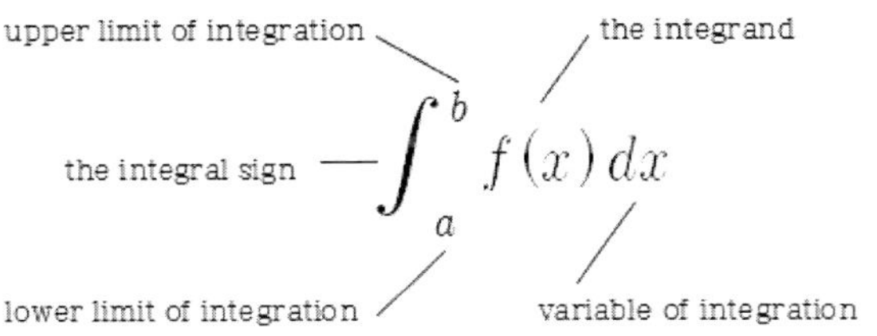

Figure 3.2: Definite Integral

- $f(x) = 1$ with the antiderivative $F(x) = x+C$, which implies that $\int f(x)dx = x+C$. If we consider $\int_a^b f(x)dx$, then it is simply the area of a square, $(b-a) \times 1 = b-a$.

- $f(x) = x$ with the antiderivative $F(x) = \frac{1}{2}x^2 + C$. If we consider $\int_a^b f(x)dx$, then it is simply the area of a trapezoid, $\frac{a+b}{2} \times (b-a) = \frac{b^2-a^2}{2}$.

- $f(x) = x^2$ with the antiderivative $F(x) = \frac{1}{3}x^3 + C$. Now, we compute $\int_0^1 f(x)dx$:

  (1) We divide the interval $[0,1]$ into n subintervals $[x_{i-1}, x_i]$, $x_i = i/n$ for $i = 0, 1, \cdots, n$.

  (2) Since $f(x) = x^2$ is increasing on $[0,1]$, m_i and M_i are respectively,

$$m_i = x_{i-1}^2 \leq f(x) \leq x_i^2 = M_i, \quad \forall x \in [x_{i-1}, x_i] .$$

  (3) Thus, since $\sum_{i=1}^n i^2 = n(n+1)(2n+1)/6$, we have

$$\begin{aligned} L_n &= \sum_{i=1}^n m_i \cdot \frac{1}{n} = \sum_{i=1}^n \left(\frac{i-1}{n}\right)^2 \frac{1}{n} = \sum_{i=1}^{n-1} \left(\frac{i}{n}\right)^2 \frac{1}{n} = \frac{(n-1)n(2n-1)/6}{n^3}, \\ U_n &= \sum_{i=1}^n M_i \cdot \frac{1}{n} = \sum_{i=1}^n \left(\frac{i}{n}\right)^2 \frac{1}{n} = \frac{n(n+1)(2n+1)/6}{n^3}. \end{aligned}$$

(4) Therefore, it follows that

$$\int_0^1 f(x)dx = \lim_{n\to\infty} L_n = \lim_{n\to\infty} U_n = \frac{1}{3}.$$

Example 3.1.3 We consider the function $f(x) : [0,1] \to \mathbb{R}$ defined by

$$f(x) = \begin{cases} 1, & \text{if } x \text{ is rational,} \\ 0, & \text{if } x \text{ is irrational.} \end{cases}$$

For instance, $f(0.5) = 1,\ f(1/\sqrt{2}) = 0$.

(1) We divide the interval $[0,1]$ into n subintervals $[x_{i-1}, x_i]$, $x_i = i/n$ for $i = 0, 1, \ldots, n$.

(2) Since each subinterval $[x_{i-1}, x_i]$ contains both rational and irrational numbers, the minimum and the maximum are $m_i = 0$ and $M_i = 1$ for all $i = 1, \ldots, n$.

(3) Thus

$$L_n = \sum_{i=1}^{n} m_i \cdot \frac{1}{n} = 0, \quad U_n = \sum_{i=1}^{n} M_i \cdot \frac{1}{n} = 1.$$

(4) Since $\lim_{n\to\infty} L_n \neq \lim_{n\to\infty} U_n$, $\int_0^1 f(x)dx$ is not defined, and f is ***not*** integrable on $[0,1]$.

Theorem 3.1.2 (Existence of Definite Integrals)

(1) Let f be a monotone function on $[a,b]$. Then f is integrable on $[a,b]$.

(2) Let f be a continuous function on $[a,b]$. Then f is integrable on $[a,b]$.

Theorem 3.1.3 (Some Properties of the Integral) *Let f and g be integrable functions on $[a,b]$ and let $k \in \mathbb{R}$. Then*

(1) $\displaystyle\int_a^b kf(x)dx = k\int_a^b f(x)dx.$

(2) $\displaystyle\int_a^b \{f(x) + g(x)\}\,dx = \int_a^b f(x)dx + \int_a^b g(x)dx.$

(3) $\displaystyle\int_a^b f(x)dx = -\int_b^a f(x)dx.$

(4) $\int_a^b f(x)dx = \int_a^c f(x)dx + \int_c^b f(x)dx.$

(5) $\int_a^b k\ dx = k(b-a)$, *for any number* k.

(6) $\int_a^a f(x)dx = 0.$

(7) If f *has a maximum value* M, *and a minimum value* m, *on* $[a, b]$, *then*

$$m \cdot (b-a) \leq \int_a^b f(x)dx \leq M \cdot (b-a).$$

(8) If $f(x) \geq g(x)$ *on* $[a, b]$, *then* $\int_a^b f(x)dx \geq \int_a^b g(x)dx.$

(9) If $f(x) > 0$ *on* $[a, b]$, *then* $\int_a^b f(x)dx \geq 0.$

Proof We only consider the proof of (2) as an illustration :

Let m_i^f be the minimum value and M_i^f be the maximum value of $f(x)$, respectively, and let m_i^g be the minimum value and M_i^g be the maximum value of $g(x)$, respectively on the interval $[x_{i-1}, x_i]$. Then,

$$m_i^f + m_i^g \leq f(x) + g(x) \leq M_i^f + M_i^g, \ \forall\ x \in [x_{i-1}, x_i].$$

Thus,

$$\lim_{n\to\infty} \sum_{i=1}^n (m_i^f + m_i^g)\frac{b-a}{n} \leq \int_a^b \{f(x) + g(x)\}\, dx \leq \lim_{n\to\infty} \sum_{i=1}^n (M_i^f + M_i^g)\frac{b-a}{n}. \quad (3.1.1)$$

In addition, in (3.1.1),

$$\text{LHS} = \lim_{n\to\infty} \sum_{i=1}^n m_i^f \frac{b-a}{n} + \lim_{n\to\infty} \sum_{i=1}^n m_i^g \frac{b-a}{n} = \int_a^b f(x)dx + \int_a^b g(x)dx.$$

and

$$\text{RHS} = \lim_{n\to\infty} \sum_{i=1}^n M_i^f \frac{b-a}{n} + \lim_{n\to\infty} \sum_{i=1}^n M_i^g \frac{b-a}{n} = \int_a^b f(x)dx + \int_a^b g(x)dx.$$

Thus, the proof of (2) is complete.

Remark 3.1.4 Suppose that f is continuous on $[a, b]$ and g is continuous on $[c, d]$, where $f([a, b]) \subset [c, d]$. Then

(1) $g \circ f$ is integrable on $[a, b]$.

(2) fg is integrable on $[a, b]$.

(3) $|f|$ is integrable on $[a, b]$ and

$$|\int_a^b f(x)dx| \leq \int_a^b |f(x)|dx.$$

Theorem 3.1.4 (Mean Value Theorem for Integrals) *If a function f is continuous on $[a, b]$, then there exists a point $c \in [a, b]$ such that*

$$\int_a^b f(x)dx = (b-a)f(c).$$

Proof By (7) of Theorem 3.1.3, we have

$$m \leq \frac{1}{b-a}\int_a^b f(x)dx \leq M,$$

where m and M are, respectively, the minimum and maximum of $f(x)$ on $[a, b]$. Since $f(x)$ is continuous, there exists a point $c \in [a, b]$ such that

$$f(c) = \frac{1}{b-a}\int_a^b f(x)dx,$$

by Intermediate Value Theorem.

Remark 3.1.5 Figure 3.3 shows that if we choose a point $c \in [a, b]$, then $\int_a^b f(x)dx$ and the rectangle with $(a, 0)$, $(a, f(c))$, $(b, f(c))$, $(b, 0)$ have the same size.

Example 3.1.4 Consider $\displaystyle\int_0^1 x^2 dx = \frac{1}{3}$. Then, there exists a point $c \in [0, 1]$ that satisfies $(1-0)c^2 = \frac{1}{3}$. That is, $c = \frac{1}{\sqrt{3}} \in [0, 1]$.

Theorem 3.1.5 (Fundamental Theorem of Calculus) *Suppose that $f(x)$ is continuous on $[a, b]$. Let $F(x) = \displaystyle\int_a^x f(t)dt$. Then*

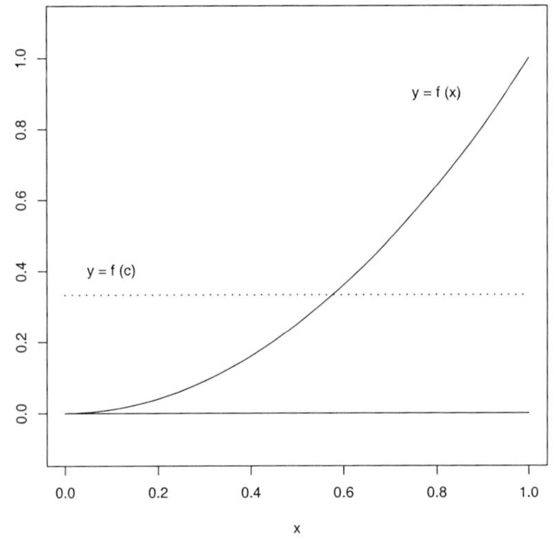

Figure 3.3: Mean Value Theorem for Integrals

(1) **Part I** *: $F(x)$ is continuous on $[a,b]$ and is an antiderivative of f. That is, F is differentiable on (a,b), and $F' = f$,*

$$\frac{d}{dx}F(x) = \frac{d}{dx}\int_a^x f(t)dt = f(x),\ x \in (a,b).$$

(2) **Part II** *:*

$$\int_a^b f(x) = G(b) - G(a) = G(x)\Big|_a^b = \Big[G(x)\Big]_a^b.$$

where G is any antiderivative of f, i.e. $G' = f$ and $G(x) = F(x) + C$, where C is an arbitrary constant.

Proof (1) **Part I** :

Suppose that $x \in [a,b]$ and $a \le x+h \le b$ for small h, then

$$\begin{aligned} F(x+h) - F(x) &= \int_a^{x+h} f(t)dt - \int_a^x f(t)dt = \int_x^{x+h} f(t)dt \\ &= h \cdot f(c), \quad \text{for some } c \in (x, x+h), \end{aligned}$$

by Theorem 3.1.4, Mean Value Theorem for integrals. Thus, we have

$$\lim_{h\to 0}\{F(x+h)-F(x)\}=0,$$

which implies that $F(x)$ is continuous.

In addition, note that for $x\in(a,b)$,

$$\frac{1}{h}\{F(x+h)-F(x)\}=f(c),\ c\in(x,x+h).$$

Therefore,

$$\lim_{h\to 0}\frac{1}{h}\{F(x+h)-F(x)\}=\lim_{h\to 0}f(c)=f(x).$$

Hence $F(x)$ is differentiable on (a,b) and $\frac{d}{dx}F(x)=f(x)$.

(2) **Part II** :

Since

$$G(x)=F(x)+C=\int_a^x f(t)dt+C,$$

that is, $G(x)$ is an indefinite integral of $f(x)$. If $x=a$, $G(a)=c$ since $F(a)=0$. Also, if $x=b$, then $G(b)=F(b)+C=\int_a^b f(t)dt+G(a)$. It follows that

$$\int_a^b f(t)dt=G(b)-G(a).$$

Remark 3.1.6 Results in Theorem 3.1.5 is known as the **fundamental theorem of calculus** . The result (2) of Theorem 3.1.5 provides a practical way to evaluate the integral of $f(x)$ on $[a,b]$.

Example 3.1.5 (1) If $F(x)=\frac{x^3}{3}$ for all $x\in[a,b]$, then $F'(x)=x^2$ for all $x\in[a,b]$. Furthermore, $f=F'$ is continuous. Therefore the Fundamental Theorem of Calculus implies that

$$\int_a^b x^2dx=F(b)-F(a)=\frac{1}{3}(b^3-a^3).$$

(2) Evaluate $\int_0^\pi \sin x dx$.

If we take $F(x)=-\cos x$, then $F'(x)=f(x)=\sin x$. Hence

$$\int_0^\pi \sin x dx=F(\pi)-F(0)=(-\cos\pi)-(-\cos 0)=1-(-1)=2.$$

(3) $\displaystyle\int_0^1 x^6 dx = \frac{x^7}{7}\bigg|_0^1 = \frac{1}{7}.$

Example 3.1.6 Compute $\displaystyle\int_{-1}^1 \frac{1}{x^2} dx.$

- If $F(x) = -\dfrac{1}{x}$, then $F'(x) = f(x) = \frac{1}{x^2}$. Therefore,

$$\int_{-1}^1 \frac{1}{x^2} dx = \left[-\frac{1}{x}\right]_{-1}^1 = -2.$$

- It is contradictory that $f(x) = \frac{1}{x^2} > 0$ but its integral is negative.

- Since $f(x) = \frac{1}{x^2}$ is not continuous at $x = 0$, we cannot apply the Fundamental Theorem of Calculus to $\displaystyle\int_{-1}^1 \frac{1}{x^2} dx.$

Example 3.1.7 Additional results about definite integrals are :

(1) $\displaystyle\int_0^b e^x dx = e^x\bigg|_0^b = e^b - 1.$

(2) $\displaystyle\int_a^b e^{-x} dx = -e^{-x}\bigg|_a^b = e^{-a} - e^{-b}.$

(4) $\displaystyle\int_0^{\frac{\pi}{2}} \cos x \; dx = \sin x\bigg|_0^{\frac{\pi}{2}} = 1.$

(5) $\displaystyle\int_a^b x^n dx = \frac{x^{n+1}}{n+1}\bigg|_a^b = \frac{1}{n+1}(b^{n+1} - a^{n+1}), \; n \neq -1.$

(6) $\displaystyle\int_a^b \frac{1}{x} dx = \log x\bigg|_a^b = \log b - \log a, \; 0 < a < b.$

(7) Even and odd functions :

(a) If f is even, i.e., $f(-x) = f(x)$, then $\displaystyle\int_{-a}^a f(x)dx = 2\int_0^a f(x)dx.$

(b) If f is odd, i.e., $f(-x) = -f(x)$, then $\displaystyle\int_{-a}^a f(x)dx = 0.$

Example 3.1.8 We can also show that $1+x \le e^x$, since $\int_a^b f(x)dx \le \int_a^b g(x)dx$ if $f(x) \le g(x)$. Since $e^x \ge 1$, if $x \ge 0$

$$\int_0^x 1\,dt \le \int_0^x e^t dt,$$

which implies that

$$x \le e^x - 1.$$

3.2 Techniques of Integration

Integration by Substitution

Theorem 3.2.1 (Substitution Theorem) *Let $J=[\alpha,\beta]$ and let $u : J \to \mathbb{R}$ have a continuous derivative on J. If $f : I \to \mathbb{R}$ is continuous on an interval I containing $g(J)$, then*

$$\int_\alpha^\beta f(g(x))g'(x)dx = \int_{g(\alpha)}^{g(\beta)} f(u)du.$$

Remark 3.2.1 If we let $u = g(x)$, then $du = g'(x)dx$. Thus, by the Chain Rule,

$$f(g(x))g'(x)dx = f(u)du.$$

Example 3.2.1 Consider the integral

(1) $\int_0^2 3x^2\sqrt{x^3+1}\,dx$. Substitute $u = x^3+1$ so that $du = 3x^2dx$. Thus,

$$\int_0^2 3x^2\sqrt{x^3+1}\,dx = \int_1^9 \sqrt{u}\,du = \left.\frac{u^{1/2+1}}{1/2+1}\right|_1^9 = \frac{52}{3}.$$

(2) $\int_0^{\log 2} e^{3x}dx$. Substitute $u = 3x$ so that $du = 3dx$. Thus,

$$\int_0^{\log 2} e^{3x}dx = \int_0^{3\log 2} e^u \frac{1}{3}dx = \left.\frac{1}{3}e^u\right|_0^{3\log 2} = \frac{7}{3}.$$

(3) $\displaystyle\int_{\pi^2/4}^{\pi^2} \frac{\sin\sqrt{t}}{\sqrt{t}}dt$. Substitute $u=\sqrt{t}$ so that $du = 1/(2\sqrt{t})dt$. Thus,

$$\int_{\pi^2/4}^{\pi^2} \frac{\sin\sqrt{t}}{\sqrt{t}}dt = \int_{\pi/2}^{\pi} \sin x dx = -2\cos x\Big|_{\pi/2}^{\pi} = 2.$$

Example 3.2.2 We can prove the identity $\log a\cdot b = \log a + \log b$ in the following way : Note that $\log x = \displaystyle\int_1^x \frac{1}{t}dt$. Thus

$$\begin{aligned}\log(a\cdot b) &= \int_1^{ab}\frac{1}{t}dt = \int_1^a \frac{1}{t}dt + \int_a^{ab}\frac{1}{t}dt \\ &= \log a + \int_1^b \frac{1}{u}du, \text{ by substituting } (t/a = u) \\ &= \log a + \log b.\end{aligned}$$

Example 3.2.3 Consider $\displaystyle\int \frac{f'(x)}{f(x)}dx$. By substituting $u = f(x)$, we have

$$\int \frac{f'(x)}{f(x)}dx = \int \frac{1}{u}du = \log|u| + C = \log|f(x)| + C.$$

For example,

$$\int_0^{\frac{\pi}{4}} \tan x dx = \int_0^{\frac{\pi}{4}} \frac{\sin x}{\cos x}dx = -\log|\cos x|\Big|_0^{\frac{\pi}{4}} = \log\sqrt{2}.$$

Example 3.2.4 Consider $\displaystyle\int_{-a}^{a} f(x)dx$.

(a) If $f(x)$ is even, i.e. $f(-x) = f(x)$, then

$$\int_{-a}^{a} f(x)dx = \int_{-a}^{0} f(x)dx + \int_0^a f(x)dx = \int_0^a f(-u)du + \int_0^a f(x)dx = 2\int_0^a f(x)dx.$$

(b) If $f(x)$ is odd, i.e. $f(-x) = -f(x)$, then

$$\int_{-a}^{a} f(x)dx = \int_{-a}^{0} f(x)dx + \int_0^a f(x)dx = \int_0^a f(-u)du + \int_0^a f(x)dx = 0.$$

(c) For example,

(i) $\displaystyle\int_{-b}^{b}(x^5+x^3+3x)dx=0.$

(ii) $\displaystyle\int_{-b}^{b} xe^{-x^2}dx=0.$

(iii) $\displaystyle\int_{-b}^{b} e^{-|x|}dx=2\int_{0}^{b} e^{-x}dx=2(1-e^{-b}).$

Example 3.2.5 Using $\displaystyle\int \frac{1}{x}dx=\log|x|+C$, we can consider the integral of fraction of two polynomials. For example

(1) $\displaystyle\int \frac{1}{x(x+2)}dx.$

$$\int \frac{1}{x(x+2)}dx=\int \frac{1}{2}\left(\frac{1}{x}-\frac{1}{x+2}\right)dx=\frac{1}{2}\log|x|-\frac{1}{2}\log|x+2|+C$$

(2) $\displaystyle\int \frac{2x+1}{x^2+3x+2}dx.$

$$\int \frac{2x+1}{x^2+3x+2}dx=\int \frac{2(x+1)-1}{(x+2)(x+1)}dx=2\int \frac{1}{x+2}dx-\int \frac{1}{(x+2)(x+1)}dx.$$

$$=2\int \frac{1}{x+2}dx-\int \frac{1}{x+1}dx+\int \frac{1}{x+2}dx=3\log|x+2|-\log|x+1|+C.$$

Example 3.2.6 (Trigonometric Integrals I)

(1) Consider $\displaystyle\int \sin^3 x\cos^2 x\,dx$. Since $\sin^2+\cos^2=1$, the integral can be rewritten as

$$\int \sin^3 x\cos^2 x\,dx=\int \sin x(1-\cos^2 x)\cos^2 x\,dx.$$

Then, using the substitution $u=\cos x$ and $du=-\sin x dx$,

$$\int \sin x(1-\cos^2 x)\cos^2 x dx=\int -(1-u^2)u^2du=-\frac{1}{4}\cos^4 x+\frac{1}{6}\cos^6 x+C.$$

(2) $\displaystyle\int \cos^3 x\,dx$ can also be calculated in the same way.

(3) $\displaystyle\int_0^{\pi/2}\sin x\cos x dx=\left[\frac{1}{2}\sin^2 x\right]_0^{\pi/2}=\frac{1}{2}.$

Example 3.2.7 (Trigonometric Integrals II) Integral by substitutions :

(1) $\displaystyle\int \frac{1}{1+x^2}dx$. Let $x = \tan\theta$. Then, $dx = \sec^2\theta d\theta$. Thus,

$$\int \frac{1}{1+x^2}dx = \int \frac{1}{1+\tan^2\theta}\sec^2\theta d\theta = \int 1\, d\theta = \theta + C = \tan^{-1}x + C.$$

(2) $\displaystyle\int \frac{1}{x^2-2x+2}dx$ can also be easily calculated by substituting $x-1=\tan\theta$.

(3) $\displaystyle\int \frac{1}{\sqrt{1-x^2}}dx$. Let $x=\sin\theta$, $-\frac{\pi}{2}<\theta<\frac{\pi}{2}$. Then $dx=\cos\theta d\theta$. Thus,

$$\int \frac{1}{\sqrt{1-x^2}}dx = \int \frac{1}{\cos\theta}\cos\theta d\theta = \theta + C = \sin^{-1}x + C.$$

(4) Consider $y=\sqrt{a^2-x^2}$, $a>0$. Then, $\displaystyle\int_{-a}^{a}\sqrt{a^2-x^2}dx = \frac{\pi a^2}{2}$ since it is the area of the top half of the circle with radius a.

Remark 3.2.2 Suppose that the function $f:\ [a,b]\to\mathbb{R}$ is continuous, and that $g(x)$ is differentiable. Then,

(1) $\displaystyle\frac{d}{dx}\left[\int_x^b f(t)dt\right] = -f(x)$.

(2) $\displaystyle\frac{d}{dx}\left[\int_a^{g(x)} f(t)dt\right] = f(g(x))g'(x)$.

Example 3.2.8 Calculate the following derivatives :

(1) $\displaystyle\frac{d}{dx}\left(\int_0^x x^2t^2dt\right)$.

(2) $\displaystyle\frac{d}{dx}\left(\int_1^{e^x} \log t dt\right)$.

(3) $\displaystyle\frac{d}{dx}\left(\int_{-x}^x e^{t^2}dt\right)$.

(4) $\displaystyle\frac{d}{dx}\left(\int_1^x \cos(x+t)dt\right)$.

Integration by Parts

Theorem 3.2.2 (Integration by Parts) *Suppose that f and g are differentiable on $[a, b]$ and that f' and g' are integrable on $[a, b]$. Then,*

$$\int_a^b f(x)g'(x)dx = \Big[f(x)g(x)\Big]_a^b - \int_a^b f'(x)g(x)dx = [f(b)g(b) - f(a)g(a)] - \int_a^b f'(x)g(x)dx.$$

Proof If $h = fg$, then $h' = fg' + f'g$ by the product rule of differentiation, and h' is also integrable on $[a, b]$. From the fundamental theorem of calculus, we then obtain $\int_a^b h'(x)dx = h(b) - h(a)$. That is,

$$\int_a^b f(x)g'(x)dx + \int_a^b f'(x)g(x)dx = f(b)g(b) - f(a)g(a).$$

Remark 3.2.3 Remember the product rule for differentiation, $(uv)' = u'v + uv'$, which can be rewritten as $uv' = (uv)' - u'v$. Thus, we have the formula for the integration by parts:

$$\int uv'dx = uv - \int u'vdx.$$

Example 3.2.9 (1) $\int_0^b xe^x dx$. Let $u = x$ and $v' = e^x$. Then, $u' = 1$ and $v = e^x$. Thus, using the integration by parts, we have

$$\int_0^b xe^x dx = xe^x\Big|_0^b - \int_0^b e^x dx = (b-1)e^b + 1.$$

(2) $\int x^2e^x dx$. Let $u = x^2$ and $v' = e^x$. Then, $u' = 2x$ and $v = e^x$. Thus,

$$\int x^2e^x dx = x^2e^x - \int 2xe^x dx = x^2e^x - 2e^x(x-1) + C.$$

(3) For $n \in \mathbb{N}$,

$$\begin{aligned} F_n &= \int x^n e^{-x} dx, \quad u = x^n,\ v' = e^{-x} \\ &= -x^n e^{-x} + \int nx^{n-1}e^{-x}dx \\ &= -x^n e^{-x} + nF_{n-1}. \end{aligned}$$

Example 3.2.10 (1) $\int \log x dx$. Let $u = \log x$ and $v' = 1$. Then, $u' = 1/x$ and $v = x$. Thus, using integration by parts we have

$$\int \log x dx = x \log x - \int 1 dx = x \log x - x + C.$$

(2) $\int x \log x dx$. Let $u = \log x$ and $v' = x$. Then, $u' = 1/x$ and $v = x^2/2$. Thus,

$$\int x \log x dx = \frac{x^2}{2} \log x - \int \frac{x^2}{2} \frac{1}{x} dx = \frac{x^2}{2} \log x - \frac{x^2}{4} + C.$$

(3) $\int x \cos x dx = x \sin x - \int \sin x dx = x \sin x + \cos x + C.$

(4) $\int e^x \cos x dx$. Let $u = \cos x$ and $v' = e^x$. Then,

$$\begin{aligned} \int e^x \cos x dx &= e^x \cos x - \int e^x (-\sin x) dx, \quad u = \sin x,\ v' = e^x \\ &= e^x \cos x + e^x \sin x - \int e^x \cos x dx. \end{aligned}$$

Therefore,

$$\int e^x \cos x dx = \frac{e^x(\cos x + \sin x)}{2} + C.$$

Example 3.2.11 Find the unique solution of the equation

$$\begin{cases} F'(x) - F(x) = x, & \text{for all } x, \\ F(0) = 2. \end{cases}$$

By multiplying e^{-x}, we have

$$\begin{cases} \frac{d}{dx} e^{-x} F(x) = e^{-x} x, & \text{for all } x, \\ F(0) = 2. \end{cases}$$

Consider

$$\begin{cases} g'(x) = e^{-x} x, & \text{for all } x, \\ g(0) = 2 \end{cases}$$

The unique solution is defined by

$$g(x) = g(0) + \int_0^x e^{-t} t dt = 3 - e^{-x}(x + 1).$$

Thus, the unique solution of $F(x)$ is given by the formula

$$F(x) = e^x(3 - e^{-x}(x + 1)), \quad \forall x.$$

Hence, $F(x) = 3e^x - x - 1$ for all x.

Example 3.2.12 (Integrating Inverses of Functions) Integration by parts leads to a rule for integrating inverses.

(1) Let $y = f^{-1}(x)$, then $x = f(y)$ and $dx = f'(y)dy$. Thus, by integration by parts with $u = y$ and $v' = f'(y)$, we have

$$\int f^{-1}(x)dx = xf^{-1}(x) - \int f(y)dy.$$

(2) Suppose that the function $f : [0, \infty) \to \mathbb{R}$ is strictly increasing and differentiable and that $f(0) = 0$. Then, we get

$$\int_0^{f(t)} f^{-1}(x)dx = xf^{-1}(x)\Big|_0^{f(t)} - \int_0^t f(y)dy.$$

Thus, we have

$$\int_0^x f(t)dt + \int_0^{f(x)} f^{-1}(t)dt = xf(x) \text{ for all } x \geq 0.$$

(3) For any number $a > 1$, we have

$$\int_1^a \log x dx + \int_0^{\log a} e^y dy = a \log a.$$

It is also easy to see the above identity using the geometric interpretation.

(4)

$$\int_0^1 \tan^{-1} x dx = x \tan^{-1} x\Big|_0^1 - \int_{\tan^{-1} 0}^{\tan^{-1} 1} \tan x dx = \tan^{-1} 1 + \log(\cos x)\Big|_{\tan^{-1} 0}^{\tan^{-1} 1}$$
$$= \frac{\pi}{4} + \log \frac{1}{\sqrt{2}}.$$

Alternatively, we may consider a direct integration by parts with $u = \tan^{-1} x$ and $v' = 1$. That is,

$$\begin{aligned} \int_0^1 \tan^{-1} x dx &= \tan^{-1} x \cdot x\Big|_0^1 - \int_0^1 \frac{1}{1+x^2} \cdot x dx \\ &= \tan^{-1} 1 - \frac{1}{2}\log(1+x^2)\Big|_0^1 = \frac{\pi}{4} - \frac{\log 2}{2}. \end{aligned}$$

3.3 Improper Integral

Throughout our discussion of the Riemann integral, we have assumed that the integrand is bounded and that the integral is being taken over a bounded interval. Now, we extend Riemann integration to the situation with the unbounded integrand and the situation with the unbounded range of integration. In such cases, the Riemann integral is called an **improper** integral. Thus, there are two kinds of improper integral. $\int_a^\infty f(x)dx$ is called an **improper integral of the first kind** , where the range of integration is infinite. On the other hand, if the integrand $f(x)$ becomes infinite inside the range of integration, then $\int_a^b f(x)dx$ is called an **improper integral of the second kind** .

Improper integral of the first kind

Definition 3.3.1 *Let f be defined on $[a, \infty)$ and integrable on $[a, c]$ for every $c > a$. If $\lim_{c\to\infty} \int_a^c f$ exists, then the improper integral of f on $[a, \infty)$, denoted by $\int_a^\infty f$, is given by*

$$\int_a^\infty f = \lim_{c\to\infty} \int_a^c f.$$

Remark 3.3.1 (1) If f is continuous on $(-\infty, b]$, then

$$\int_{-\infty}^b f(x)dx = \lim_{a\to-\infty} \int_a^b f(x)dx.$$

(2) If the limit exists, the integral **converges** (finite). If the limit does not exist, the integral **diverges** (infinite).

Example 3.3.1 (1) Let $f(x) = x^{-2}$ for $x \geq 1$. Then,

$$\int_1^\infty \frac{1}{x^2}dx = \lim_{b\to\infty} \int_1^b \frac{1}{x^2}dx = \lim_{b\to\infty} \left[-\frac{1}{x}\right]_1^b = \lim_{b\to\infty} (-\frac{1}{b} + 1) = 1.$$

(2) Consider $\int_1^\infty e^{-x}dx$. Then

$$\begin{aligned} \int_1^\infty e^{-x}dx &= \lim_{N\to\infty} \int_1^N e^{-x}dx \\ &= \lim_{N\to\infty} \left[-e^{-x}\right]_1^N = \lim_{N\to\infty} (-e^{-N} + e^{-1}) = e^{-1}. \end{aligned}$$

(3) Let $g(x) = \cos x$ for $x \geq 0$. Then for each $c > 0$ we have

$$\int_0^c \cos x dx = \sin c - \sin 0 = \sin c.$$

Since $\lim_{c\to\infty} \sin c$ does not exist, $\int_0^\infty \cos x dx$ has no meaning.

Theorem 3.3.1 (Comparison Test) *Let $f(x)$ be a nonnegative function that is Riemann integrable on $[a, b]$. Suppose that there exists a function $g(x)$ such that $f(x) \leq g(x)$ for $x \geq a$. If $\int_a^\infty g(x)dx$ converges, then so does $\int_a^\infty f(x)dx$ and we have*

$$\int_a^\infty f(x)dx \leq \int_a^\infty g(x)dx.$$

Theorem 3.3.2 (Limit Comparison Test) *Let $f(x)$ and $g(x)$ be nonnegative functions that are Riemann integrable on $[a, b]$. If*

$$\lim_{x\to\infty} \frac{f(x)}{g(x)} = k,$$

where k is a positive constant, then $\int_a^\infty f(x)dx$ and $\int_a^\infty g(x)dx$ are either both convergent or both divergent.

Theorem 3.3.3 (Direct Comparison Test) *Suppose that $f(x)$ and $g(x)$ be nonnegative on $[a, \infty)$.*

(1) If $f(x) \leq g(x)$ as $x \to \infty$ and $\int_a^\infty g(x)dx < \infty$, then $\int_a^\infty f(x)dx < \infty$.

(2) If $f(x) \geq g(x)$ as $x \to \infty$ and $\int_a^\infty g(x)dx$ is diverging, then $\int_a^\infty f(x)dx$ is diverging.

Example 3.3.2 For $x \geq 1$, we have

$$e^x \leq e^{x^2},$$

and thus

$$e^{-x} \geq e^{-x^2}.$$

Therefore, by Theorem 3.3.3 $\int_1^\infty e^{-x}dx < \infty$ implies that $\int_1^\infty e^{-x^2}dx < \infty$.

Example 3.3.3 For what values of p does the integral $\displaystyle\int_1^\infty \frac{1}{x^p}dx$ converge? That is, when does the integral converge, and what is the value of integral?

Suppose that $p \neq 1$. Then,

$$\begin{aligned}\int_1^\infty \frac{1}{x^p}dx &= \lim_{N\to\infty}\int_1^N \frac{1}{x^p}dx = \lim_{N\to\infty}\left[\frac{x^{-p+1}}{-p+1}\right]_1^N \\ &= \lim_{N\to\infty}\frac{1}{1-p}\left(N^{-p+1}-1\right) = \lim_{N\to\infty}\frac{1}{1-p}\left(\frac{1}{N^{p-1}}-1\right) \\ &= \begin{cases}\text{converges to } \frac{1}{p-1}, & \text{if } p>1, \\ \text{diverges}, & \text{if } p<1.\end{cases}\end{aligned}$$

If $p = 1$, the integral diverges, since

$$\int_1^\infty \frac{1}{x}dx = \lim_{N\to\infty}\int_1^N \frac{1}{x}dx = \lim_{N\to\infty}\left[\log x\right]_1^N = \lim_{N\to\infty}(\log N - \log 1) = \infty.$$

Therefore, $\displaystyle\int_1^\infty \frac{1}{x^p}dx$ converges when $p > 1$ and diverges when $0 < p \leq 1$.

Example 3.3.4 Investigate the convergence of the following integrals.

(1) $\displaystyle\int_1^\infty \frac{\log x}{x}dx.$

By integration by parts with $u = \log x$, $v' = \frac{1}{x}$, we have

$$\begin{aligned}\int_1^N \frac{\log x}{x}dx &= \left[(\log x)^2\right]_1^N - \int_1^N \frac{\log}{x}dx \\ \int_1^N \frac{\log x}{x}dx &= \frac{1}{2}(\log N)^2.\end{aligned}$$

Therefore, $\displaystyle\int_1^\infty \frac{\log x}{x}dx = \lim_{N\to\infty}\int_1^N \frac{\log x}{x}dx$ diverges.

(2) $\displaystyle\int_1^\infty \frac{\log x}{x^2}dx.$

By integration by parts with $u = \log x$, $v' = \frac{1}{x^2}$, we have

$$\begin{aligned}\lim_{N\to\infty}\int_1^N \frac{\log x}{x^2}dx &= \lim_{N\to\infty}\left[(\log x)\left(-\frac{1}{x}\right)\right]_1^N - \lim_{N\to\infty}\int_1^N\left(-\frac{1}{x^2}dx\right) \\ &= \lim_{N\to\infty}\int_1^N \frac{1}{x^2}dx = 1.\end{aligned}$$

Therefore, $\displaystyle\int_1^\infty \frac{\log x}{x^2}dx = 1.$

(3) $\displaystyle\int_1^\infty \frac{1}{1+x^2}dx.$

By making the substitution of $x = \tan\theta$, we have

$$\begin{aligned}\int_1^\infty \frac{1}{1+x^2}dx &= \int_{\frac{\pi}{4}}^{\frac{\pi}{2}} \frac{\sec^2\theta}{1+\tan^2\theta}d\theta \\ &= \frac{\pi}{4}.\end{aligned}$$

Thus, $\int_1^\infty \frac{1}{1+x^2}dx$ converges and equals $\frac{\pi}{4}$.

Example 3.3.5 Indeed, it is easy to see that $\displaystyle\int_1^\infty \frac{1}{1+x^2}dx$ converges by Direct Comparison Test as well as Limit Comparison Test.

First, since $\displaystyle\int_1^\infty \frac{1}{x^2}dx < \infty$ and $1/x^2 \geq 1/(1+x^2)$, it follows that $\int_1^\infty \frac{1}{1+x^2}dx$ converges by Direct Comparison Test. Also, let $f(x) = \frac{1}{x^2}$ and $g(x) = \frac{1}{1+x^2}$. Then,

$$\lim_{x\to\infty} \frac{\frac{1}{x^2}}{\frac{1}{1+x^2}} = \lim_{x\to\infty} \frac{1+x^2}{x^2} = 1.$$

Since $\displaystyle\int_1^\infty \frac{1}{x^2}dx$ is finite, $\int_1^\infty \frac{1}{1+x^2}$ is also finite by Limit Comparison Test.

Remark 3.3.2 Note that

(1) $\displaystyle\int_0^\infty \frac{1}{1+x^2}dx = \int_0^1 \frac{1}{1+x^2}dx + \int_1^\infty \frac{1}{1+x^2}dx < \infty$. It is clear that the first integral is finite since the range is finite as well as the integrand is bounded on $[0,1]$, and the second integral is finite by comparison test. Indeed,

$$\int_0^\infty \frac{1}{1+x^2}dx = \tan^{-1}x\Big|_{x=0}^{\infty} = \frac{\pi}{2}.$$

(2) $\displaystyle\int_{-\infty}^\infty \frac{1}{1+x^2}dx = \int_{-\infty}^0 \frac{1}{1+x^2}dx + \int_0^\infty \frac{1}{1+x^2}dx = 2\int_0^\infty \frac{1}{1+x^2}dx = \pi < \infty.$

(3) In fact, $\dfrac{1}{\pi(1+x^2)}$ is a Cauchy density function.

Example 3.3.6 The integral $\displaystyle\int_0^\infty \frac{\sin x}{(x+1)^2}dx$ is convergent.

Since

$$\frac{|\sin x|}{(x+1)^2} \le \frac{1}{(x+1)^2}$$

and

$$\int_0^\infty \frac{1}{(x+1)^2}dx = \left[-\frac{1}{x+1}\right]_0^\infty = 1.$$

Definition 3.3.2 Improper Integrals of Type I *can be generalized as follows :*

(1) If $f(x)$ is continuous on $[a,\infty)$, then

$$\int_a^\infty f(x)dx = \lim_{b\to\infty}\int_a^b f(x)dx.$$

(2) If $f(x)$ is continuous on $(-\infty, b]$, then

$$\int_{-\infty}^b f(x)dx = \lim_{a\to-\infty}\int_a^b f(x)dx.$$

(3) If $f(x)$ is continuous on $(-\infty,\infty)$, then

$$\int_{-\infty}^\infty f(x)dx = \int_{-\infty}^c f(x)dx + \int_c^\infty f(x)dx,$$

where c is any real number.

In each case, if the limit is finite we say that the improper integral **converges** *and that the limit is the value of the improper integral. If the limit fails to exist, the improper integral* **diverges**.

Example 3.3.7 $\displaystyle\int_{-\infty}^\infty f(x)dx$ may not be equal to $\displaystyle\lim_{b\to\infty}\int_{-b}^b f(x)dx$.

(1) Consider $\int_{-\infty}^{\infty} e^{-x^2} dx$.

Since $e^{-x^2} < e^{-x}$ when $x > 1$, we have

$$\int_0^\infty e^{-x^2} dx = \int_0^1 e^{-x^2} + \int_1^\infty e^{-x^2} dx < 1 + \int_1^\infty e^{-x} dx = 1 + e^{-1} < \infty.$$

Thus,

$$\int_{-\infty}^{\infty} e^{-x^2} dx = \int_{-\infty}^{0} e^{-x^2} dx + \int_0^\infty e^{-x^2} dx = 2\int_0^\infty e^{-x^2} dx.$$

Also, note that xe^{-x^2} is an odd function and that $\int_0^\infty xe^{-x^2} dx < \infty$. Therefore, we have

$$\int_{-\infty}^{\infty} xe^{-x^2} dx = 0.$$

(2) Consider $\int_{-\infty}^{\infty} \frac{x}{1+x^2} dx$.

Note that $\int_1^\infty \frac{x}{1+x^2} dx$ diverges, since

$$\int_1^\infty \frac{x}{1+x^2} dx = \lim_{N\to\infty} \int_1^\infty \frac{x}{1+x^2} dx = \lim_{N\to\infty} \frac{1}{2} \log(1+x^2)\Big|_1^N = \infty$$

Thus $\int_0^\infty \frac{x}{1+x^2} dx$ also diverges, since

$$\int_0^\infty \frac{x}{1+x^2} dx = \int_0^1 \frac{x}{1+x^2} dx + \int_1^\infty \frac{x}{1+x^2} dx \ge \int_1^\infty \frac{x}{1+x^2} dx.$$

Therefore, although $\frac{x}{1+x^2}$ is an odd function, $\int_{-\infty}^{\infty} \frac{x}{1+x^2} dx$ is not defined. Indeed, it is not equal to $\lim_{b\to\infty} \int_{-b}^{b} \frac{x}{1+x^2} dx = 0$. However, notice that $\int_{-\infty}^{\infty} \frac{1}{1+x^2} dx = \pi < \infty$.

Gamma Function

Definition 3.3.3 *The function* $\Gamma : \mathbb{R}^+ \to \mathbb{R}$ *defined by*

$$\Gamma(x) = \int_0^\infty t^{x-1} e^{-t} dt, \quad x > 0$$

is called the **gamma function** .

Remark 3.3.3 The gamma function is defined by an improper integral that converges for all $x > 0$. To see that, we break up the integral at 1,

$$\int_0^\infty t^{x-1}e^{-t}dt = \int_0^1 t^{x-1}e^{-t}dt + \int_1^\infty t^{x-1}e^{-t}dt.$$

- If $x > 1$, $t^{x-1}e^{-t}$ is bounded for all $t \in [0, 1]$.
- If $0 < x < 1$, then $t^{x-1}e^{-t} < t^{x-1}$, and $\int_0^1 t^{x-1}dt$ converges.
- Since $\lim_{t\to\infty} t^{x+1}e^{-t} = 0$, there exists a real number $M > 1$ such that $t^{x+1}e^{-t} \le 1$ for all $t \ge M$. Therefore,

$$\int_1^\infty t^{x-1}e^{-t}dt = \int_1^M t^{x-1}e^{-t}dt + \int_M^\infty t^{x-1}e^{-t}dt < \int_1^M t^{x-1}e^{-t}dt + \int_M^\infty \frac{1}{t^2}dt < \infty.$$

Hence, the gamma function Γ is well defined.

Example 3.3.8 Gamma functions : $\Gamma(x) = \displaystyle\int_0^\infty t^{x-1}e^{-t}dt$,

(1) If $x = 1$, $\displaystyle\int_0^\infty e^{-x}dx = \int_0^1 e^{-x}dx + \int_1^\infty e^{-x}dx < \infty$. Indeed,

$$\int_0^\infty e^{-x}dx = e^{-x}\Big|_0^\infty = 1.$$

(2) If $x = 2$, $\displaystyle\int_0^\infty xe^{-x}dx = -xe^{-x}\Big|_0^\infty + \int_0^\infty e^{-x}dx = 1.$

(3) If $x = 3$, $\displaystyle\int_0^\infty x^2e^{-x}dx = -x^2e^{-x}\Big|_0^\infty + \int_0^\infty 2xe^{-x}dx = 2.$

Theorem 3.3.4 *If $x > 0$, then $\Gamma(x+1) = x\Gamma(x)$. Therefore, if $n \in \mathbb{N}$, then $\Gamma(n+1) = n!$.*

Proof Since $\Gamma(x)$ converges, integration by parts yields

$$\Gamma(x+1) = \int_0^\infty t^x e^{-t}dt = -t^x e^{-t}\Big|_0^\infty + \int_0^\infty xt^{x-1}e^{-t}dt = x\,\Gamma(x).$$

Note that $\Gamma(1) = \int_1^\infty e^{-t}dt = 1$. Therefore, it follows that $\Gamma(n+1) = n\Gamma(n) = n(n-1)\cdots\Gamma(1) = n!$.

Remark 3.3.4 If $x, y > 0$, then $\dfrac{\Gamma(x)\Gamma(y)}{\Gamma(x+y)}$ defines a new function called the **beta function** , $B(x, y)$, which is a function of two variables x and y.

Improper integral of the second kind

Definition 3.3.4 *Let f be defined on $(a,b]$ and integrable on $[c,b]$ for every $c \in (a,b]$. If $\lim_{c\to a+}\int_c^b f$ exists, then the* **improper integral** *of f on $(a,b]$, denoted by $\int_a^b f$, is given by*

$$\int_a^b f = \lim_{c\to a+}\int_c^b f.$$

If f is not defined at $x=b$ and continuous on $[a,b)$, then

$$\int_a^b f(x)dx = \lim_{c\to b^-}\int_a^c f(x)dx.$$

Remark 3.3.5 Integrals of functions that become infinite at a point within the interval of integration arc **improper integrals of Type 2** . We call the point where the given function is not defined *singularity point* .

Theorem 3.3.5 *Let $f(x)$ be a nonnegative function such that $\int_c^b f(x)dx$ exists for every c in $(a,b]$. If there exists a function $g(x)$ such that $f(x)\le g(x)$ for all x in $(a,b]$, and if $\int_c^b g(x)dx$ converges as $c\to a^+$, then so does $\int_c^b f(x)dx$ and we have*

$$\int_a^b f(x)dx \le \int_a^b g(x)dx.$$

Theorem 3.3.6 Let $f(x)$ and $g(x)$ be nonnegative functions that are Riemann integrable on $[c,b]$ for every c such that $a<c\le b$. If

$$\lim_{x\to a^+}\frac{f(x)}{g(x)} = k,$$

where k is a positive constant, then $\int_a^b f(x)dx$ and $\int_a^b g(x)dx$ are either both convergent or both divergent.

Example 3.3.9 Investigate the convergence of the following integrals.

(1) $\int_0^1 \frac{1}{x}dx.$

$$\int_0^1 \frac{1}{x}dx = \lim_{a\to 0+}\int_a^1 \frac{1}{x}dx = \lim_{a\to 0+}\log x|_a^1 = \lim_{a\to 0+}(-\log a).$$

Since the limit does not exist, the improper integral diverges.

(2) $\int_0^1 \frac{1}{x^2}dx$.

$$\int_0^1 \frac{1}{x^2}dx = \lim_{a\to 0^+}\int_a^1 \frac{1}{x^2}dx = \lim_{a\to 0^+} -\frac{1}{x}\Big|_a^1 = \lim_{a\to 0^+}\left(-1+\frac{1}{a}\right).$$

Since the limit does not exist, the improper integral diverges.

(3) $\int_0^1 \frac{1}{\sqrt{x}}dx$.

$$\int_0^1 \frac{1}{\sqrt{x}}dx = \lim_{c\to 0^+}\int_c^1 \frac{1}{\sqrt{x}}dx = \lim_{c\to 0^+}[2\sqrt{x}]_c^1 = \lim_{c\to 0^+}[2-2\sqrt{c}] = 2.$$

Since the limit exists, the improper integral converges.

Example 3.3.10 If f is not defined or discontinuous at $x = c$, an interior point of $[a, b]$, then

$$\int_a^b f(x)dx = \int_a^c f(x)dx + \int_c^b f(x)dx, \quad \text{for } a < c < b.$$

(1) $\int_{-1}^1 \frac{1}{\sqrt{|x|}}dx = \int_{-1}^0 \frac{1}{\sqrt{|x|}}dx + \int_0^1 \frac{1}{\sqrt{x}}dx = 2\int_0^1 \frac{1}{\sqrt{x}} = 4.$

(2) $\int_{-1}^1 \frac{1}{x^2}dx = \int_{-1}^0 \frac{1}{x^2}dx + \int_0^1 \frac{1}{x^2}dx.$

However, note that $\int_0^1 \frac{1}{x^2}dx$ diverges. Same is true for $\int_{-1}^0 \frac{1}{x^2}dx$.

To extend this, if f is not defined or discontinuous at $x = c_1, \cdots, c_n$ (n interior points) on $[a, b]$, (Here, $a < c_1 < \cdots < c_n < b, n = 1, 2, 3, \cdots$)

$$\int_a^b f(x)dx = \int_a^{c_1} f(x)dx + \int_{c_1}^{c_2} f(x)dx + \cdots + \int_{c_n}^b f(x)dx.$$

Example 3.3.11 Consider $\int_0^1 \frac{1}{x^p}dx$. Then, for $p \neq 1$,

$$\int_0^1 \frac{1}{x^p}dx = \lim_{a\to 0^+}\frac{x^{-p+1}}{-p+1}\Big|_a^1 = \frac{1}{1-p}\lim_{a\to 0^+}(1-a^{1-p}).$$

Thus, $\int_0^1 \frac{1}{x^p}dx$

- converges if $1-p>0$.
- diverges if $1-p\le 0$.

Note that $\int_0^1 \frac{1}{x^p}dx = \int_1^\infty \frac{1}{u^{2-p}}du$ and that the latter integral converges if $2-p>1$ and diverges if $2-p\le 1$.

3.4 Numerical Integration

The integration of a continuous function f defined in $[a,b]$, $\int_a^b f(x)dx = S$ is the area between the curve $y=f(x)$ and $y=0$, when $a\le x\le b$. If the integrand f is too complicated to find an antiderivative, we resort to numerical integration. Numerical integration yields numbers rather than analytic expressions. For instance, there is no formula for $\int_a^b \exp(-x^2)dx$. We will consider three techniques for numerical integration: Riemann sums, the Trapezoidal Rule, and Simpson's Rule.

(1) The **Riemann sum** approximate $\int_a^b f(x)dx$ by the sum of rectangular areas:

$$\int_a^b f(x)dx \approx \sum_{i=1}^n y_{i-1}\Delta_x,\quad \Delta_x = x_i - x_{i-1},\ \text{i.e. } x_i = x_{i-1}+\Delta_x,$$

where $a=x_0<x_1<x_2<\cdots<x_n=b$ and $y_0=f(x_0)$, $y_1=f(x_1),\ldots,y_n=f(x_n)$. See Figure 3.4.

1. The **Trapezoidal Rule** divides up the area below the function into trapezoids instead of rectangles (See Figure 3.5). Note that the area of a trapezoid is the average height times horizontal width. Thus,

$$\begin{aligned}
&\text{Total Trapezoidal Area in Figure 3.5}\\
=\ &\Delta_x\left(\frac{y_0+y_1}{2}+\frac{y_1+y_2}{2}+\cdots+\frac{y_{n-1}+y_n}{2}\right) = \sum_{i=1}^n \frac{\Delta_x}{2}(y_{i-1}+y_i)\\
=\ &\frac{\Delta_x}{2}(y_0+2y_1+2y_2+\cdots+2y_{n-1}+y_n) \approx \int_a^b f(x)dx,
\end{aligned}$$

where $\Delta_x=(b-a)/n$, $x_0=a$, $x_1=a+\Delta_x$, $x_2=a+2\Delta_x,\ldots,x_{n-1}=a+(n-1)\Delta_x$, $x_n=b$, $y_i=f(x_i)$, $i=0,1,\ldots,n$.

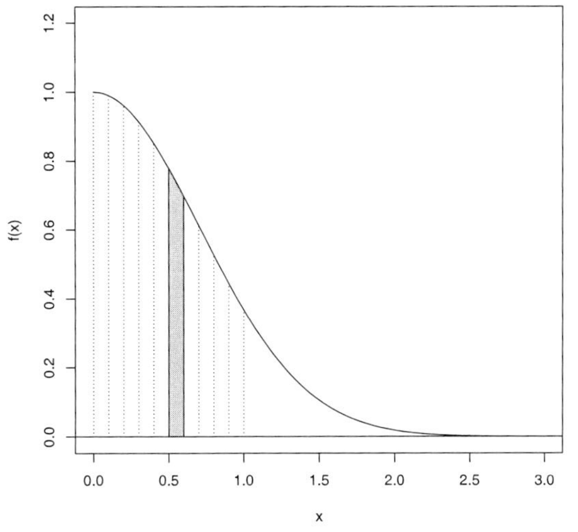

Figure 3.4: Riemann sum for $f = \exp(-x^2)$ on $[a, b]$

2. The **Simpson's Rule** , named after the mathematician Thomas Simpson, is to use the quadratic polynomials, i.e. the curve with parabolas for approximating the integral as in Figure 3.6. That is, to approximate $\int_a^b f(x)dx$, use

$$\int_a^b f(x)dx \approx \frac{\Delta_x}{3}(y_0 + 4y_1 + 2y_2 + 4y_3 + \cdots + 2y_{n-2} + y_n).$$

Note that the curve $y = Ax^2 + Bx + C$ in Figure 3.6 passes through three points, $(-h, y_0)$, $(0, y_1)$, and (h, y_2), where $y_0 = Ah^2 - Bh + C$, $y_1 = C$ and $y_2 = Ah^2 + Bh + C$. By solving these equations, we get

$$A = \frac{1}{2h^2}(y_0 + y_2 - 2y_1),\ B = \frac{1}{2h}(y_2 - y_0),\ C = y_1.$$

Thus, the shaded area $(= S)$ is given by $\int_{-h}^{h}(Ax^2 + Bx + C)dx = \frac{2}{3}Ah^3 + 2Ch$. In other words,

$$S = \frac{h}{3}(y_0 + y_2 - 2y_1) + 2hy_1 = \frac{h}{3}(y_0 + 4y_1 + y_2).$$

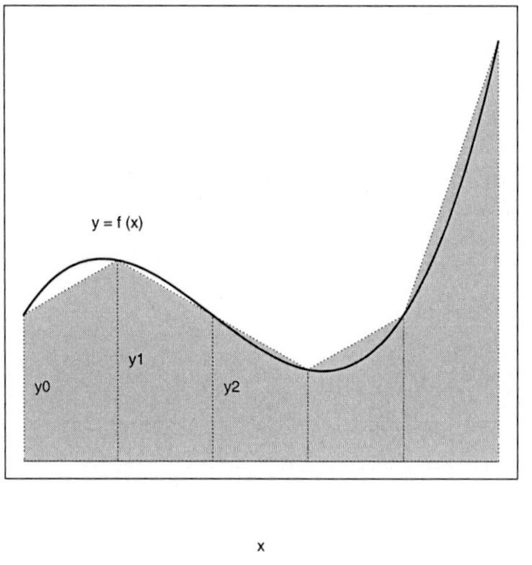

Figure 3.5: Trapezoidal Rule approximates $\int_a^b f(x)dx$ by the area of trapezoids.

Example 3.4.1 We can evaluate following integrals by numerical methods.

(1) $\int_0^1 \frac{1}{\sqrt{2\pi}} e^{-\frac{x^2}{2}} dx (= I)$. For $n = 10$, Trapezoidal Rule yields $I = 0.341143$ and Simpson's Rule yields $I = 0.341345$. The integral is actually the same as the probability that a standard normal random variable is between 0 and 1, which is calculculated as 0.3413447 by a statistical package, R.

(2) $\int_0^1 \frac{1}{\pi(1+x^2)} dx (= I)$. Consider three points $(0, 1/\pi)$, $(1/2, 4/5\pi)$ and $(1, 1/2\pi)$. By the Trapezoidal Rule, we get

$$\int_0^1 \frac{1}{\pi(1+x^2)}) dx \approx \Delta_x \left(\frac{y_0}{2} + y_1 + \frac{y_2}{2} \right) = \frac{1}{2} \left(\frac{1}{2\pi} + \frac{4}{5\pi} + \frac{1}{4\pi} \right) = 0.2467.$$

By Simpson's Rule, we get

$$\frac{\Delta_x}{3} (y_0 + 4y_1 + y_2) = \frac{1/2}{3} \left(\frac{1}{\pi} + 4\frac{4}{5\pi} + \frac{1}{2\pi} \right) = 0.2493.$$

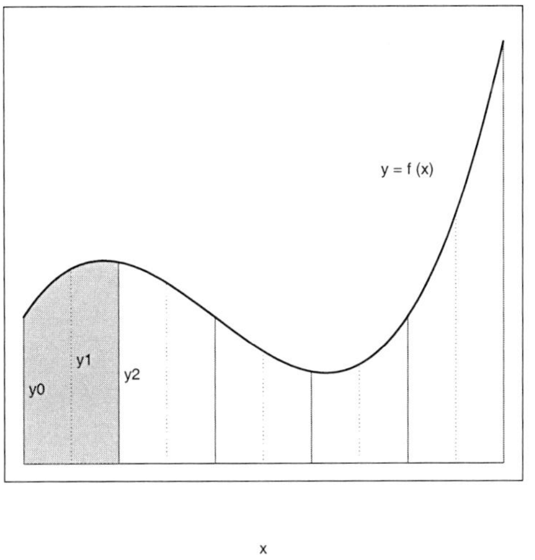

Figure 3.6: Simpson's Rule approximates $\int_a^b f(x)dx$ by the integral of quadratic functions.

The integral is actually the same as the probability that a Cauchy random variable with parameters 0 and 1 is between 0 and 1, which is calculculated as 0.25 by a statistical package, R. Note that the exact answer is given by

$$\int_0^1 \frac{1}{\pi(1+x^2)} dx = \left.\frac{\tan^{-1} x}{\pi}\right|_0^1 = \frac{1}{\pi}\left(\frac{\pi}{4} - 0\right) = 0.25.$$

3.5 Suggested Exercises

1. Find the limits.

   (a)
   $$\lim_{n\to\infty}\left(\frac{1}{n+1}+\frac{1}{n+2}+\cdots+\frac{1}{2n}\right).$$

   (b)
   $$\lim_{n\to\infty}\frac{1}{n}\left(e^{1/n}+e^{2/n}+\cdots+e^{(n-1)/n}+e^{n/n}\right).$$

2. Suppose that x and y are related by the equation
   $$x=\int_0^y\frac{1}{\sqrt{1+4t^2}}dt.$$
   Show that d^2y/dx^2 is proportional to y and find the constant of proportionality.

3. If f is a continuous function, find the value of the integral
   $$I=\int_0^a\frac{f(x)}{f(x)+f(a-x)}dx,$$
   by using the substitution.

4. Find $f(3)$ and $f'(2)$ if $f(x)=e^{g(x)}$ and $g(x)=\displaystyle\int_2^x\frac{2t}{1+t^2}dt.$

5. Prove that if $f(x)$ is a continuous function, then
   $$\int_0^a f(x)dx=\int_0^a f(a-x)dx.$$

6. Evaluate
   $$\lim_{x\to3}\left(\frac{x}{x-3}\int_3^x\frac{\sin t}{t}dt\right).$$

7. Find $f(4)$ if $\displaystyle\int_0^x f(t)dt=x\cos\pi x.$

8. Evaluate the following integrals.

   (a) $\displaystyle\int(\cos x)e^{\sin x}dx.$

(b) $\displaystyle\int \frac{1}{1+e^x}dx.$

(c) $\displaystyle\int_2^4 x^{2x}(1+\log x)\,dx.$

9. Use integration by parts to establish the reduction formula.

(a) $\displaystyle\int x^n e^{ax}dx = \frac{x^n e^{ax}}{a} - \frac{n}{a}\int x^{n-1}e^{ax}dx,\ a \neq 0.$

(b) $\displaystyle\int (\log x)^n dx = x(\log x)^n - n\int (\log x)^{n-1}dx.$

10. Evaluate the following integrals.

(a) $\displaystyle\int \frac{e^{4t}+2e^{2t}-e^t}{e^{2t}+1}dt.$

(b) $\displaystyle\int \frac{\sin\theta}{\cos^2\theta+\cos\theta-2}d\theta.$

11. Suppose that a function $f(x)$ satisifes the following conditions,

$$f(1) = \frac{1}{3}, \quad \int f(x)dx = xf(x) + \frac{1}{x^2}.$$

Determine $f(x)$.

12. Let $F(x) = \int_0^x xe^{t^2}dt$ for $x \in [0,1]$. Find $F''(x)$ for $x \in (0,1)$.

13. Evaluate $\displaystyle\lim_{x\to 0}\frac{\int_0^x \sqrt{3+t^2}dt}{x}$.

14. Let f be continuous on $[0,\infty)$. Suppose that $f(x) \neq 0$ for all $x > 0$ and that $[f(x)]^2 = 2\int_0^x f(t)dt$ for all $x \geq 0$. Determine $f(x)$ for all $x \geq 0$.

15. Suppose that the function $f : \mathbb{R} \to \mathbb{R}$ has a continuous second derivative. Fix a number a. Prove that

$$\int_a^x f''(t)(x-t)dt = -(x-a)f'(a) + f(x) - f(a), \quad \text{for all } x.$$

16. Suppose that the function $f : \mathbb{R} \to \mathbb{R}$ has a continuous second derivative. Prove that for any two numbers a and b,

$$\int_a^b xf''(x)x = bf'(b) + f(a) - af'(a) - f(b).$$

17. Evaluate

(a) $\displaystyle\int_0^\infty x^2 e^{-x} dx.$

(b) $\displaystyle\int_{-1}^1 3x^5\sqrt{x^3+1}dx.$

18. Evaluate the following integrals.

(a) $\displaystyle\int \frac{1}{x\log x} dx.$

(b) $\displaystyle\int \frac{x+4}{x^2+5x-6} dx.$

19. Investigate the following integrals if they converge or diverge

(a) $\displaystyle\int_1^\infty \frac{e^{-t}}{\sqrt{t}} dt.$

(b) $\displaystyle\int_{-\infty}^\infty \frac{1}{x^2(1+e^x)} dx.$

20. Show that

(a) $\displaystyle\int_0^\infty \frac{\log x}{1+x^2} dx$ converges.

(b) $\displaystyle\int_0^\infty \frac{\log x}{1+x^2} dx = 0.$ (Hint : Use the substitution $u = 1/x$)

21. Investigate the convergence of the following integrals.

(a) $\displaystyle\int_0^2 \frac{1}{1-x^2} dx.$

(b) $\displaystyle\int_0^2 \frac{1}{1-x} dx.$

(c) $\displaystyle\int_{-\infty}^\infty \frac{1}{e^x+e^{-x}} dx.$

22. Evaluate $\displaystyle\int_{-\infty}^0 xe^x dx.$

23. Investigate the convergence or divergence of $\displaystyle\int_1^\infty \frac{1+e^{-x}}{x} dx.$

24. For what value or values of a does

$$\int_1^\infty \left(\frac{ax}{x^2+1} - \frac{1}{2x} \right) dx$$

converge? Evaluate the corresponding integral(s).

25. Suppose that the function $f : [0, \pi/2] \to \mathbb{R}$ is continuous and strictly increasing and that $f : [0, \pi/2] \to \mathbb{R}$ is differentiable. Consider the formula

$$\int_a^x f(t)dt + \int_{f(a)}^{f(x)} f^{-1}(t)dt = xf(x) - af(a), \quad 0 \le a \le x \le \pi/2.$$

   (a) Prove this formula.

   (b) Calculate $\int_0^{e^{\pi/2}} g^{-1}(x)dx =$, where $g(x) = e^x \sin x$.

26. Suppose that the function $f : [0, \infty) \to [0, \infty)$ is continuous and strictly increasing and that $f : [0, \infty) \to [0, \infty)$ is differentiable. Moreover, assume $f(0) = 0$. We would like to prove the following inequality, known as *Young's inequality*.

$$ab \le F(a) + G(b) \quad \text{for all } a \ge 0, \text{ and } b \ge 0,$$

where

$$F(x) = \int_0^x f(t)dt \text{ and } G(x) = \int_0^x f^{-1}(t)dt, \quad \forall x \ge 0.$$

   (a) Provide a geometric interpretation of this formula in terms of areas.

   (b) Suppose that $b = f(a)$. Then show that the equality holds, i.e.

$$af(a) = F(a) + G(b).$$

   (c) Suppose that $b \ge f(a)$. Then establish *Young's inequality*, i.e.

$$ab \le F(a) + G(b), \quad \forall a \ge 0, \text{ and } b \ge f(a).$$

   Similarly, we can show that *Young's inequality* also holds when $b \le f(a)$.

27. If $f(x)$ is continuous for $x \geq 0$, the **Laplace transform** of f is the function $\psi(t)$ defined by

$$\psi(t) = \int_0^\infty f(x)e^{-tx}dt$$

and the domain of ψ is the set consisting of all numbers t for which the integral converges. Find the Laplace transform of the following functions.

(a) $f(x) = 1$.

(b) $f(x) = e^x$.

(c) $f(x) = xe^x$.

28. Show that $\displaystyle\int_0^\infty x^2e^{-x^2}dx = \frac{1}{2}\int_0^\infty e^{-x^2}dx$.

29. Show that if $\alpha > -1$ and $\beta > \alpha + 1$, then the following integral is convergent.

$$\int_0^\infty \frac{x^\alpha}{1+x^\beta}dx.$$

Chapter 4

Infinite Series

In this chapter, we discuss the convergence properties of infinite series of real numbers and develop several useful tests for determining whether a given series is convergent or divergent. In addition, we consider power series and Taylor series at the end.

4.1 Convergence of Infinite Series

Definition 4.1.1 *Given a sequence $\{a_n\}$, we have the sequence $\{s_n\}$: $s_n = \sum_{k=1}^{n} a_k$. The following object*

$$a_1 + a_2 + a_3 + \cdots \quad \text{or} \quad \sum_{n=1}^{\infty} a_n$$

is called an **infinite series** *or a* **series** *. The numbers s_n are called the* ***partial sums*** *of the series. If $\{s_n\}$ converges to s, we say that* **the series converges** *and write:*

$$\lim_{n\to\infty} s_n = \sum_{n=1}^{\infty} a_n = s.$$

We say the series $\sum_{n=1}^{\infty} a_n$ **diverges** *if $\{s_n\}$ diverges.*

Example 4.1.1 (Telescoping series): Let $\{a_n\}$ and $\{b_n\}$ be two sequences such that $a_n = b_n - b_{n-1}$ for $n = 1, 2, ...$ and $b_0 = 0$. Then $\sum a_n$ converges if, and only if, $\lim_{n\to\infty} b_n$ exists, in which case we have

$$\sum_{n=1}^{\infty} a_n = \lim_{n\to\infty} b_n$$

For the infinite series $\sum_{n=1}^{\infty} \frac{1}{n(n+1)}$, we have the partial sums given by

$$\begin{aligned} s_n &= \frac{1}{1\cdot 2} + \frac{1}{2\cdot 3} + \cdots + \frac{1}{n(n+1)} \\ &= \left(\frac{1}{1} - \frac{1}{2}\right) + \left(\frac{1}{2} - \frac{1}{3}\right) + \cdots + \left(\frac{1}{n} - \frac{1}{n+1}\right) \\ &= 1 - \frac{1}{n+1}. \end{aligned}$$

Since the sequence $\{s_n\}$ of partial sums converges to 1, we have $\sum_{n=1}^{\infty} \frac{1}{n(n+1)} = 1$.

Remark 4.1.1 Convergence of a series is formulated in terms of convergence of the associated sequence of its partial sums.

(1) If $\sum_{n=1}^{\infty} a_n$ is a convergent series, then $\lim_{n\to\infty} a_n = 0$. However, the converse is not true. For examples, $\sum_{n=1}^{\infty}(\sqrt{n+1} - \sqrt{n})$ and $\sum_{n=1}^{\infty} \frac{1}{n}$ are divergent, and their nth terms go to zero as $n \to \infty$. On the other hand, if $\lim_{n\to\infty} a_n \neq 0$, then $\sum_{n=1}^{\infty} a_n$ is divergent.

(2) A monotone sequence of real numbers is convergent if and only if it is bounded. Thus, the infinite series of nonnegative terms, $\sum a_n$, $a_n \geq 0$ is convergent if and only if it is bounded.

Example 4.1.2 (Harmonic Series) The harmonic series $\sum_{n=1}^{\infty} 1/n$ has the partial sums

$$s_n = 1 + \frac{1}{2} + \frac{1}{3} + \cdots + \frac{1}{n}.$$

Then

$$\begin{aligned} s_n &= 1 + \frac{1}{2} + \frac{1}{3} + \frac{1}{4} + \frac{1}{5} + \frac{1}{6} + \frac{1}{7} + \frac{1}{8} + \frac{1}{9} + \cdots \\ s_{2^k} &= 1 + \frac{1}{2} + \left(\frac{1}{3} + \frac{1}{4}\right) + \left(\frac{1}{5} + \frac{1}{6} + \frac{1}{7} + \frac{1}{8}\right) + \cdots \\ &> 1 + \frac{1}{2} + \frac{1}{2} + \cdots + \frac{1}{2} = 1 + \frac{1}{2}k \end{aligned}$$

Thus, the sequence $\{s_n\}$ is increasing and not bounded above. Therefore, it can not be convergent.

Theorem 4.1.1 (Geometric Series)

(a) If $|r| < 1$*, then* $\sum_{n=0}^{\infty} ar^n$ *is convergent, with sum* $\frac{a}{1-r}$.

(b) If $|r| \geq 1$ *and* $a \neq 0$*, then* $\sum_{n=0}^{\infty} ar^n$ *is divergent.*

Convergence Tests

Theorem 4.1.2 (Comparison Test) *Let* $\sum_{n=1}^{\infty} a_n$ *and* $\sum_{n=1}^{\infty} b_n$ *be infinite series of nonnegative terms. Then*

(1) If $0 \leq a_n \leq b_n$*, for* $n = 1, 2, \ldots$*, and* $\sum_{n=1}^{\infty} b_n$ *is convergent, then* $\sum_{n=1}^{\infty} a_n$ *is convergent.*

(2) If $0 \leq b_n \leq a_n$*, for* $n = 1, 2, \ldots$*, and* $\sum_{n=1}^{\infty} b_n$ *is divergent, then* $\sum_{n=1}^{\infty} a_n$ *is divergent.*

Example 4.1.3 (1) Consider the series $\sum 1/(n+1)^2$. Then, we have

$$0 < \frac{1}{(n+1)^2} < \frac{1}{n(n+1)}$$

In Example 4.1.1, $\sum \frac{1}{n(n+1)}$ was shown to be convergent, so *Comparison Test* implies that $\sum \frac{1}{(n+1)^2}$ also converges.

(2) $\sum \frac{1}{n^3}$ is convergent, since $\frac{1}{n^3} \leq \frac{1}{n^2}$ and $\sum \frac{1}{n^2}$ is convergent.

(3) The series $\sum \frac{1}{\sqrt{n}}$ is divergent.

(4) Consider $\sum_{n=1}^{\infty} \frac{1}{n!}$. Let $a_n = \frac{1}{n!}$ and $b_n = \frac{1}{2^{n-1}}$. Then $a_n \leq b_n$. Since $\sum_{n=1}^{\infty} \frac{1}{2^{n-1}} < \infty$, $\sum_{n=1}^{\infty} \frac{1}{n!} < \infty$. Therefore, $\sum_{n=1}^{\infty} \frac{1}{n!}$ converges by Comparison Test.

Theorem 4.1.3 (Limit Comparison Test) *Suppose that* $\sum_{n=1}^{\infty} a_n$ *and* $\sum_{n=1}^{\infty} b_n$ *be infinite series of nonnegative terms.*

(1) *If* $\lim_{n\to\infty} \frac{a_n}{b_n} = L,\ L \neq 0,$ *then* $\sum a_n$ *and* $\sum b_n$ *both converge or both diverge.*

(2) *If* $\lim_{n\to\infty} \frac{a_n}{b_n} = 0$ *and* $\sum b_n$ *converges, then* $\sum a_n$ *converges.*

(3) *If* $\lim_{n\to\infty} \frac{a_n}{b_n} = \infty$ *and* $\sum b_n$ *diverges, then* $\sum a_n$ *diverges.*

Remark 4.1.2 (1) Alternatively, suppose that there exists N such that for all $n \geq N$, $a_n \leq b_n$. Then, if $\sum_{n=1}^{\infty} b_n$ converges, then $\sum_{n=1}^{\infty} a_n$ converges.

(2) If $\sum_{n=1}^{\infty} a_n$ converges, then $\lim_{n\to\infty} a_n = 0$. This can be weakened to the following statement. If $\sum_{n=1}^{\infty} a_n$ converges, then $\lim_{N\to\infty} \sum_{n=N}^{\infty} a_n = 0$. The converse of the first statement is not true, but the converse of second one is true.

Example 4.1.4 In order to investigate the convergence of $\displaystyle\sum_{n=1}^{\infty} \frac{\log n}{n^3}$, consider $\displaystyle\sum_{n=1}^{\infty} \frac{n^{1/2}}{n^3}$, which is convergent. Since $\displaystyle\lim_{n\to\infty} \frac{\frac{\log n}{n^3}}{\frac{n^{1/2}}{n^3}} = 0$, by the Comparison Test, we see that $\displaystyle\sum_{n=1}^{\infty} \frac{\log n}{n^3}$ converges.

Theorem 4.1.4 (Integral Test) *Let f be a continuous function defined on $[1, \infty)$, and suppose that f is positive and decreasing. That is, if $x_1 < x_2$, then $f(x_1) \geq f(x_2) > 0$. Then the series $\displaystyle\sum_{n=1}^{\infty} f(n)$ converges if and only if*

$$\lim_{n\to\infty} \left(\int_1^n f(x)dx \right)$$

exists.

Proof Since f is decreasing, we have $f(n+1) \leq f(x) \leq f(n)$ for all $x \in [n, n+1]$. It follows that

$$f(n+1) \leq \int_n^{n+1} f(x)dx \leq f(n).$$

Geometrically, this means that the area under the curve $y = f(x)$ from $x = n$ to $x = n+1$ is between $f(n+1)$ and $f(n)$ as in Figure 4.1.

Thus, if we let $a_n = f(n)$ and $b_n = \int_n^{n+1} f(x)dx$, then by the Comparison Test, $\sum a_n$ converges if and only if $\sum b_n$ converges. Note that $\sum b_n$ converges when $\lim_{n\to\infty} \int_1^n f(x)dx$ exists.

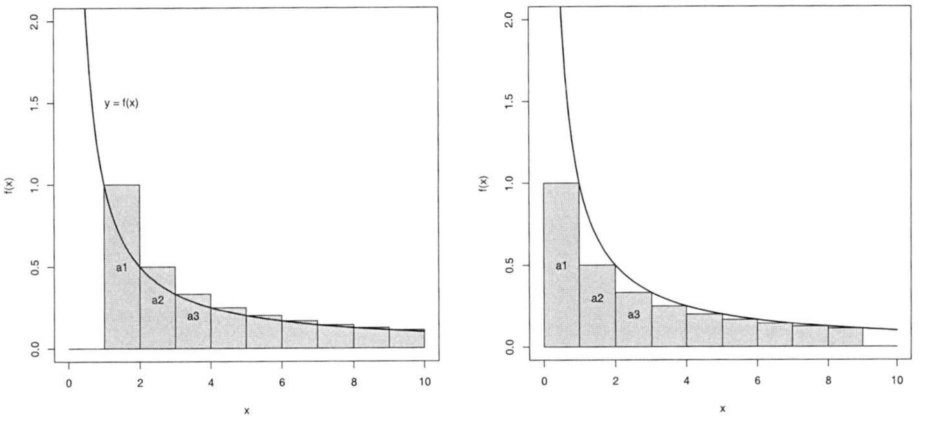

Figure 4.1: The integral test: each rectangular area can be either $f(n+1)$ or $f(n)$.

Example 4.1.5 Consider the p-series, $\sum 1/n^p$. Let $f(x) = 1/x^p$. Then,

$$\int_1^n \frac{1}{x^p} dx = \begin{cases} \frac{1}{1-p}\left(n^{1-p} - 1\right), & p \neq 1, \\ \log n - \log 1, & p = 1. \end{cases}$$

The limit of this as $n \to \infty$ will be finite if $p > 1$ and infinite if $p \leq 1$. Thus by the integral test $\sum 1/n^p$ converges if $p > 1$ and diverges if $p \leq 1$.

Example 4.1.6 It is easy to see that infinite series $\sum_{n=1}^{\infty} \frac{1}{1+n^2}$ converges by Comparison Test or Integral Test.

Theorem 4.1.5 (Ratio Test) *Let $\sum a_n$ be a series of positive terms and suppose that*

$$\lim_{n \to \infty} \frac{a_{n+1}}{a_n} = \rho.$$

(1) If $0 \leq \rho < 1$, then the series converges.

(2) If $\rho > 1$, then the series diverges.

(3) If $\rho = 1$, then the test is inconclusive.

Remark 4.1.3 If $\rho = 1$, then the test gives us no information on whether the series converges. For example, if $a_n = \frac{1}{n}$, then $\rho = 1$. We have seen that the series $\sum a_n$ diverges. On the other hand, if $a_n = \frac{1}{n^2}$, then $\rho = 1$ but $\sum a_n$ converges.

Example 4.1.7 We can use the Ratio Test to determine the convergence of the following series :

(1) $\displaystyle\sum_{n=1}^{\infty} \frac{n}{2^n}$.

(2) $\displaystyle\sum_{n=1}^{\infty} \frac{1}{n!}$.

(3) $\displaystyle\sum_{n=1}^{\infty} \frac{3^n}{n!}$.

(4) $\displaystyle\sum_{n=1}^{\infty} \frac{x^n}{n!}, \quad x > 0$.

(4) $\displaystyle\sum_{n=1}^{\infty} \frac{(2n)!}{n! \cdot n!}$ diverges.

Theorem 4.1.6 (Root Test) *Given a series $\sum a_n$, let $\rho = \lim_{n\to\infty} a_n^{1/n}$.*

(1) If $\rho < 1$, then the series converges.

(2) If $\rho > 1$, then the series diverges.

(3) If $\rho = 1$, the test is inconclusive, i.e. it does not give any information about convergence.

Example 4.1.8 Consider the series

(1) $\sum n/2^n$. Letting $a_n = 2^n$, we have $\rho = 1/2$. Thus $\sum n/2^n$ converges.

(2) For the series $\sum 1/n$, the root test fails.

Definition 4.1.2 *A series $\sum_{n=1}^{\infty} a_n$ is* **absolutely convergent** *if $\sum_{n=1}^{\infty} |a_n|$ is convergent. A series $\sum_{n=1}^{\infty} a_n$ that is convergent but not absolutely convergent is* **conditionally convergent**.

Example 4.1.9 The series $\sum_{n=1}^{\infty} \frac{(-1)^{n+1}}{n^2}$ is absolutely convergent but $\displaystyle\sum_{n=1}^{\infty} \frac{(-1)^{n+1}}{n}$ is not absolutely convergent since $\displaystyle\sum_{n=1}^{\infty} \frac{1}{n}$ is divergent. However, $\displaystyle\sum_{n=1}^{\infty} \frac{(-1)^{n+1}}{n}$ is conditionally convergent.

Theorem 4.1.7 (Alternating Series) *If $\{a_n\}$ is a decreasing sequence of positive numbers and $\lim a_n = 0$, then the series $\displaystyle\sum_{n=1}^{\infty}(-1)^{n+1}a_n$ converges. That is, the series*

$$\sum_{n=1}^{\infty}(-1)^{n+1}a_n = a_1 - a_2 + a_3 - a_4 + \cdots$$

converges if the following conditions are satisfied:

(1) $a_n \geq 0$ (The a_n's are all nonnegative).

(2) a_n is decreasing ($a_n \geq a_{n+1}$) for all $n \geq N$, for some integer N.

(3) $\lim_{n\to\infty} a_n = 0$.

Example 4.1.10 The alternating harmonic series

$$\sum_{n=1}^{\infty}(-1)^{n+1}\frac{1}{n} = 1 - \frac{1}{2} + \frac{1}{3} - \frac{1}{4} + \cdots$$

satisfies the three conditions in Theorem 4.1.7 with $N = 1$. Therefore, the series converges.

Remark 4.1.4 (Absolute and Conditional Convergence)

(1) $\sum_{n=1}^{\infty} a_n$ converges absolutely (is absolutely convergent) if $\sum_{n=1}^{\infty} |a_n|$ converges.

(2) $\sum_{n=1}^{\infty} a_n$ converges conditionally if $\sum_{n=1}^{\infty} a_n$ converges but $\sum_{n=1}^{\infty} |a_n|$ diverges.

(3) If $\sum_{n=1}^{\infty} a_n$ converges absolutely, then $\sum_{n=1}^{\infty} a_n$ converges.

Note that $\sum_{n=1}^{\infty} a_n$ converges if $\sum_{n=1}^{\infty} |a_n|$ converges. Let $a_n = a_n + |a_n| - |a_n|$. Then, $\sum_{n=1}^{\infty} a_n = \sum_{n=1}^{\infty} \{a_n + |a_n|\} - \sum_{n=1}^{\infty} |a_n|$. Since $\sum_{n=1}^{\infty} \{a_n + |a_n|\} \leq 2\sum_{n=1}^{\infty} |a_n| < \infty$, $\sum_{n=1}^{\infty} a_n$ converges.

Example 4.1.11 For $\sum_{n=1}^{\infty}(-1)^{n+1}\frac{1}{n^2} = 1 - \frac{1}{4} + \frac{1}{9} - \frac{1}{16} + \cdots$, the corresponding series of absolute values is the convergent series

$$\sum_{n=1}^{\infty}\frac{1}{n^2} = 1 + \frac{1}{4} + \frac{1}{9} + \frac{1}{16} + \cdots .$$

The original series converges because it converges absolutely.

Example 4.1.12 Consider

$$\sum_{n=0}^{\infty}\frac{x^n}{n!} = 1 + x + \frac{x^2}{2} + \frac{x^3}{3!} + \cdots (= e^x).$$

Let $a_n = \dfrac{x^n}{n!}$. Then, $\left|\dfrac{a_{n+1}}{a_n}\right| = \dfrac{|x|}{n+1} \to 0$ as $n \to \infty$. Thus, $\sum_{n=0}^{\infty}\dfrac{x^n}{n!}$ converges absolutely for all x. Therefore, $\sum_{n=0}^{\infty}\dfrac{x^n}{n!}$ converges for all x.

4.2 Power Series

Definition 4.2.1 *Let $\{a_n\}_{n=0}^{\infty}$ be a sequence of real numbers. The series*

$$\sum_{n=0}^{\infty} a_n x^n = a_0 + a_1 x + a_2 x^2 + a_3 x^3 + \cdots$$

is called a **power series** . *The number a_n is called the nth* **coefficient** *of the series.*

Remark 4.2.1 A power series about $x = 0$ is a series of the form

$$\sum_{n=0}^{\infty} a_n x^n = a_0 + a_1 x + a_2 x^2 + \cdots + a_n x^n + \cdots .$$

A power series about $x = c$ is a series of the form

$$\sum_{n=0}^{\infty} a_n (x-c)^n = a_0 + a_1(x-c) + a_2(x-c)^2 + \cdots + a_n(x-c)^n + \cdots ,$$

in which the center c and the coefficients a_0, a_1, a_2, $\cdots$, a_n, $\cdots$ are constants.

Definition 4.2.2 *If there exists a number $\rho > 0$ such that $\sum_{n=0}^{\infty} a_n x^n$ is convergent if $|x| < \rho$ and is divergent if $|x| > \rho$, then ρ is said to be the* **radius of convergence** *of the series, and the interval $(-\rho, \rho)$ is called the* **interval of convergence**.

Theorem 4.2.1 *Let ρ be the radius of convergence of $\sum_{n=0}^{\infty} a_n x^n$. Suppose that $\rho > 0$. Then $\sum_{n=0}^{\infty} a_n x^n$ converges absolutely for all x inside the interval $(-\rho, \rho)$.*

Example 4.2.1 Consider the power series whose coefficients are all equal to 1,

$$\sum_{n=0}^{\infty} x^n = 1 + x + x^2 + \cdots + x^n + \cdots$$

This geometric series converges (to $\frac{1}{1-x}$) if and only if $|x| < 1$. In other words, it can be written as

$$\frac{1}{1-x} = 1 + x + x^2 + \cdots + x^n + \cdots, \quad -1 < x < 1.$$

Theorem 4.2.2 *Let $\sum_{n=0}^{\infty} a_n x^n$ be a power series. Suppose that*

$$\lim_{n \to \infty} \left| \frac{a_{n+1}}{a_n} \right| = r.$$

Then the radius of convergence of the power series is

$$\rho = \begin{cases} 1/r, & 0 < r < \infty, \\ 0, & r = \infty, \\ \infty, & r = 0. \end{cases}$$

Theorem 4.2.3 *Let $\sum_{n=0}^{\infty} a_n x^n$ be a power series. Suppose that*

$$\lim_{n \to \infty} |a_n|^{1/n} = r.$$

Then the radius of convergence of the power series is

$$\rho = \begin{cases} 1/r, & 0 < r < \infty, \\ 0, & r = \infty, \\ \infty, & r = 0. \end{cases}$$

Example 4.2.2 For what values of x does $\sum_{n=1}^{\infty} \frac{x^n}{n!}$ converge?

Since $|\frac{a_{n+1}}{a_n}| = |\frac{n!}{(n+1)!}| = \frac{1}{n+1} \to 0$, the radius of convergence is ∞. Thus, the series converges absolutely for all x.

Example 4.2.3 Consider the series

$$\frac{1}{1 \cdot 3}x^2 + \frac{1}{2 \cdot 3^2}x^4 + \frac{1}{3 \cdot 3^3}x^6 + \frac{1}{4 \cdot 3^4}x^8 + \cdots = \sum_{n=1}^{\infty} \frac{1}{n \cdot 3^n}x^{2n}.$$

We can apply the ratio test to the series $\lim_{n \to \infty} \left| \frac{a_{n+1}}{a_n} \right| = \frac{1}{3}$. Thus, the series converges when $|x^2| < 3$ or $-\sqrt{3} < x < \sqrt{3}$. However, if the series is given by $\sum_{n=1}^{\infty} \frac{1}{n3^n}x^n$, the interval of convergence is given by $[-3, 3)$.

Term by Term Differentiation

If $\sum a_n(x-c)^n$ converges for $c - \rho < x < c + \rho$ for some $\rho > 0$, it defines a function f:

$$f(x) = \sum_{n=0}^{\infty} a_n(x-c)^n, \; c - \rho < x < c + \rho.$$

Such a function f has derivatives of all orders inside the interval of convergence. We can obtain the derivatives by differentiating the original series term by term:

$$\begin{aligned} f'(x) &= \sum_{n=1}^{\infty} na_n(x-c)^{n-1} \\ f''(x) &= \sum_{n=2}^{\infty} n(n-1)a_n(x-c)^{n-2}, \text{ and so on} \end{aligned}$$

Each of these derived series converges at every interior point of the interval of convergence of the original series.

Example 4.2.4 Consider

$$f(x) = \frac{1}{1-x} = 1 + x + x^2 + \cdots + x^n + \cdots = \sum_{n=0}^{\infty} x^n, \; -1 < x < 1.$$

We can find $f'(x)$ and $f''(x)$ by term by term differentiation.

$$\begin{aligned} f'(x) &= \frac{1}{(1-x)^2} = 1 + 2x + 3x^2 + \cdots + nx^{n-1} + \cdots \\ &= \sum_{n=1}^{\infty} nx^{n-1}, \quad -1 < x < 1, \\ f''(x) &= \frac{2}{(1-x)^3} = 2 + 6x + 12x^2 + \cdots + n(n-1)x^{n-2} + \cdots \\ &= \sum_{n=2}^{\infty} n(n-1)x^{n-2}, \quad -1 < x < 1. \end{aligned}$$

Term by Term Integration

Suppose that $f(x) = \sum_{n=0}^{\infty} a_n(x-c)^n$ converges for $c - \rho < x < c + \rho \quad (\rho > 0)$. Then $\displaystyle\sum_{n=0}^{\infty} a_n \frac{(x-c)^{n+1}}{n+1}$ converges for $c - \rho < x < c + \rho$ and

$$\int f(x)dx = \sum_{n=0}^{\infty} a_n \frac{(x-c)^{n+1}}{n+1} + C$$

for $c - \rho < x < c + \rho$.

Example 4.2.5 The series $\dfrac{1}{1+x} = 1 - x + x^2 - x^3 + \cdots$ converges on the open interval $-1 < x < 1$. Therefore, using term by term integration, we have

$$\begin{aligned} \log(1+x) &= \int_0^x \frac{1}{1+t} dt = \left[t - \frac{t^2}{2} + \frac{t^3}{3} - \frac{t^4}{4} + \cdots \right]_0^x \\ &= x - \frac{x^2}{2} + \frac{x^3}{3} - \frac{x^4}{4} + \cdots, \quad -1 < x < 1. \end{aligned}$$

4.3 Taylor Series

Definition 4.3.1 *Let f be n-times differentiable on an open interval containing the point a.* **Taylor polynomial of degree** n *for f at a is the polynomial*

$$P_n(x) = f(a) + \frac{f'(a)}{1!}(x-a) + \frac{f''(a)}{2!}(x-a)^2 + \cdots + \frac{f^{(n)}(a)}{n!}(x-a)^n.$$

Remark 4.3.1 Note that

$$P_n(a) = f(a),\ P_n'(a) = f'(a), \ldots, P_n^{(n)}(a) = f^{(n)}(a).$$

That is, f and P_n have the same value at a and have equal derivative at a of all orders up to and including n.

Example 4.3.1 Consider $f(x) = \sin x$. We easily determine the Taylor polynomial $P_3(x)$ for $\sin x$ at $a = 0$ and $a = \pi/2$. Note that

$$f(0) = 0,\ f(\pi/2) = 1,\ f'(0) = 1,\ f'(\pi/2) = 0,$$

$$f''(0) = 0,\ f''(\pi/2) = -1,\ f^{(3)}(0) = -1,\ f^{(3)}(\pi/2) = 0.$$

Thus, $P_3(x) = x - \dfrac{x^3}{3!}$ for $a = 0$, and $P_3(x) = 1 - \dfrac{1}{2}\left(x - \dfrac{\pi}{2}\right)^2$ for $a = \pi/2$.

Example 4.3.2 The function $f(x) = \frac{1}{1-x}$.

By repeated differentiation it is easy to verify that, for $k = 1, 2, \ldots$

$$f^{(k)}(x) = \frac{k!}{(1-x)^{k+1}}.$$

Thus the Taylor polynomial of degree n for f at 0 is

$$P_n(x) = \sum_{k=0}^{n} \frac{f^{(k)}(0)}{k!} x^k = 1 + x + x^2 + \cdots + x^n.$$

Note that if $|x| < 1$, then $P_n(x) \to f(x)$ as $n \to \infty$.

Example 4.3.3 Find the Taylor polynomial of degree 4 for $f(x) = e^x$ at 0. Since

$$\begin{aligned} f'(x) &= f''(x) = \ldots = f^{(4)}(x) = e^x, \\ f'(0) &= f''(0) = \ldots = f^{(4)}(0) = 1, \\ e^x &\approx P_4(x) = 1 + x + x^2/2 + x^3/6 + x^4/24 \end{aligned}$$

Figure 4.2 shows the graphs of $y = e^x$ and its first and second order Taylor polynomial. For example, we get $e^{0.5}$ by calculator, which is 1.6487. Also,

$$\begin{aligned} 1 + 0.5 &= 1.5 \\ 1 + 0.5 + 0.5^2/2 &= 1.6250 \\ 1 + 0.5 + 0.5^2/2 + 0.5^3/6 &= 1.6458 \\ 1 + 0.5 + 0.5^2/2 + 0.5^3/6 + 0.5^4/24 &= 1.6484 \end{aligned}$$

we can see it approaches 1.6487.

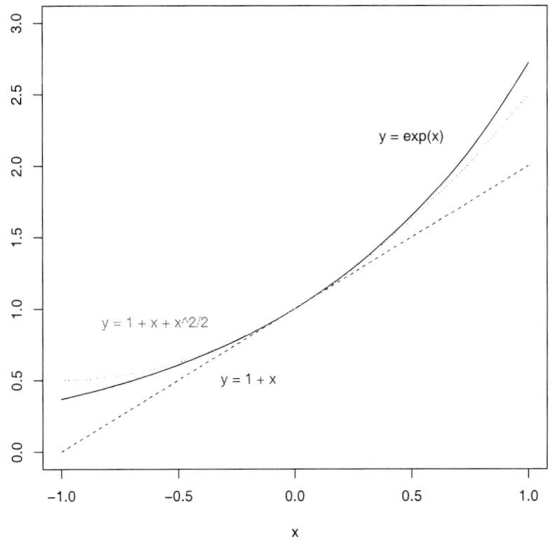

Figure 4.2: The graphs of $y = e^x$ and its first and second Taylor polynomial.

Example 4.3.4 Consider Taylor polynomial of order 7 for $f(x) = \sin x$ at 0.

$$\sin x \approx P_7(x) = x - \frac{x^3}{3!} + \frac{x^5}{5!} - \frac{x^7}{7!}.$$

Figure 4.3 shows the graphs of $y = \sin x$ and its first and second order Taylor expansion.

Definition 4.3.2 *Let f be a function with derivatives of all orders on an open interval containing the points a and x. Then the* **Taylor series** *generated by f at $x = a$ is*

$$\sum_{n=0}^{\infty} \frac{f^{(n)}(a)}{n!}(x-a)^n = f(a) + f'(a)(x-a) + \frac{f''(a)}{2!}(x-a)^2 + \cdots$$

Remark 4.3.2 We call this power series the **Taylor series** for f at a. A special form of Taylor's series is **Maclaurin's series** when $a = 0$.

Theorem 4.3.1 (Taylor's Theorem) *Let f be $(n+1)$-times differentiable on an open interval containing the points a and x. Then*

$$f(x) = f(a) + f'(a)(x-a) + \frac{f''(a)}{2!}(x-a)^2 + \cdots + \frac{f^{(n)}(a)}{n!}(x-a)^n + R_n(x),$$

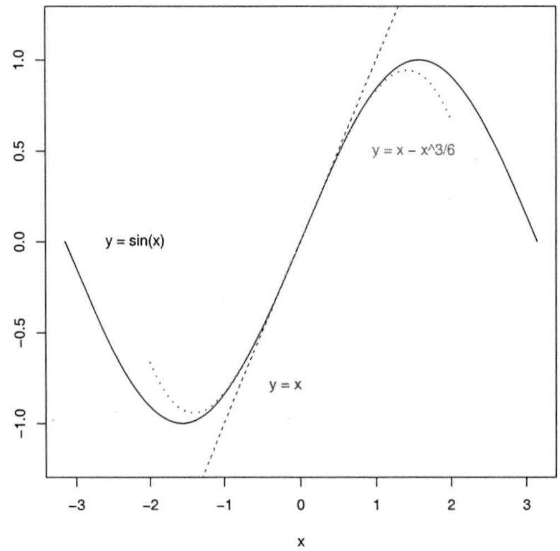

Figure 4.3: The graphs of $y = \sin x$ and its first and second Taylor polynomial.

where $R_n(x) = \dfrac{f^{(n+1)}(c)}{(n+1)!}(x-a)^{n+1}$, *and* c *is some point between* a *and* x.

Remark 4.3.3 (1) When $n = 0$, Taylor's Theorem reduces to the Mean Value Theorem.

(2) $f(x) = P_n(x) + R_n(x)$, and $R_n(x)$ is called the *remainder term* or *error term.*

(3) If $\left|f^{(n)}(c)\right| \le M$ for all c between a and x, then

$$|R_n(x)| \le \frac{M}{(n+1)!}|x-a|^{n+1}.$$

(4) It is often referred to as **Lagrange Remainder Theorem.**

Theorem 4.3.2 (Taylor Series for f at a) *Let f have derivatives of all orders on an open interval containing the points a and x. If $R_n(x) \to 0$ as $n \to \infty$, then*

$$f(x) = \sum_{n=0}^{\infty} \frac{f^{(n)}(a)}{n!}(x-a)^n. \tag{4.3.1}$$

Remark 4.3.4 The formula (4.3.1) is called a **Taylor series expansion** of the function f about the point a. (4.3.1) holds at x if and only if

$$\lim_{n\to\infty}[f(x) - P_n(x)] = 0.$$

Example 4.3.5 Find the Taylor polynomial of degree n and the Taylor series generated by $f(x) = e^x$ at $x = 0$.

Since $f(x) = e^x$, $f'(x) = e^x, \cdots, f^{(n)}(x) = e^x, \cdots$, we have $f(0) = e^0 = 1$, $f'(0) = 1, \cdots$, $f^{(n)}(0) = 1$, $\cdots$. Therefore, the Taylor polynomial of order n at $x = 0$ is

$$P_n(x) = 1 + x + \frac{x^2}{2} + \cdots + \frac{x^n}{n!},$$

and the Taylor series generated by f at $x = 0$ is

$$\begin{aligned} &f(0) + f'(0)x + \frac{f''(0)x^2}{2!} + \cdots + \frac{f^{(n)(0)}}{n!}x^n + \cdots \\ &= 1 + x + \frac{x^2}{2} + \cdots + \frac{x^n}{n!} + \cdots = \sum_{n=0}^{\infty}\frac{x^n}{n!}. \end{aligned}$$

Example 4.3.6 Let $f(x) = \log(1 + x)$ for $x > -1$. The nth Taylor polynomial for $f : (-1, \infty) \to \mathbb{R}$ at $x = 0$ is given by

$$P_n(x) = x - \frac{x^2}{2} + \cdots + (-1)^{n+1}\frac{x^n}{n},$$

and the remainder estimate is given by

$$R_n(x) = (-1)^{n+2}\frac{x^{n+1}}{(n+1)(1+c)^{n+1}}, \quad |c| \le |x|.$$

Example 4.3.7 (Basic Taylor series)

(1) $\displaystyle\frac{1}{1-x} = 1 + x + x^2 + x^3 + \cdots = \sum_{n=0}^{\infty} x^n$, for $|x| < 1$.

(2) $\displaystyle\log(1 + x) = x - \frac{x^2}{2} + \frac{x^3}{3} - \cdots = \sum_{n=1}^{\infty}(-1)^{n+1}\frac{x^n}{n}$, for $|x| < 1$.

(3) $e^x = 1 + x + \dfrac{x^2}{2!} + \dfrac{x^3}{3!} + \cdots = \displaystyle\sum_{n=0}^{\infty} \frac{x^n}{n!}$, for $x \in \mathbb{R}$.

(4) $\sin x = x - \dfrac{x^3}{3!} + \dfrac{x^5}{5!} - \cdots = \displaystyle\sum_{n=0}^{\infty} (-1)^n \frac{x^{2n+1}}{(2n+1)!}$, for $x \in \mathbb{R}$.

(5) $\cos x = 1 - \dfrac{x^2}{2!} + \dfrac{x^4}{4!} - \cdots = \displaystyle\sum_{n=0}^{\infty} (-1)^n \frac{x^{2n}}{(2n)!}$, for $x \in \mathbb{R}$.

For (4), let $f(x) = \sin x$. Then, by Taylor's Theorem, we have $\sin x = P_n(x) + R_n(x)$, where $R_n(x) = \dfrac{f^{(n+1)}(c)}{(n+1)!} x^{n+1}$, where c is between 0 and x. Note that $f^{(n+1)}(c) = \pm \sin c$ or $\pm \cos c$, implying that $|f^{(n+1)}(c)| \le 1$. Thus, it follows that $R_n(x) \to 0$ as $n \to \infty$.

Example 4.3.8 We also use Taylor's Theorem to prove the limit involving e,

$$\lim_{n\to\infty} \left(1 + \frac{1}{n}\right)^n = e.$$

(1) First calculate $P_1(x)$ and $R_1(x)$ for the function $f(x) = \log(1+x)$ at 0, i.e. for $|x| < 1$,

$$\log(1+x) = f(0) + f'(0)x + \frac{f''(c)}{2!} x^2 = x - \frac{x^2}{2(1+c)^2} \approx x.$$

(2) Then, for $c_n \in (0, 1/n)$, we have

$$\log\left(1 + \frac{1}{n}\right) = \frac{1}{n} - \frac{1}{2n^2(1+c_n)^2} \approx \frac{1}{n},$$

and

$$n \log\left(1 + \frac{1}{n}\right) = 1 - \frac{1}{2n(1+c_n)^2} \approx 1.$$

Therefore, we get to the conclusion

$$\lim_{n\to\infty} \left(1 + \frac{1}{n}\right)^n = e.$$

Example 4.3.9 For any $k \in \mathbb{N}$, and for all $x > 0$, we will show that

$$x - \frac{1}{2}x^2 + \cdots - \frac{1}{2k} x^{2k} < \log(1+x) < x - \frac{1}{2}x^2 + \cdots + \frac{1}{2k+1} x^{2k+1}.$$

Since the derivative of $\log(1+x)$ is $1/(1+x)$, the Taylor polynomial at 0 is

$$P_n(x) = x - \frac{1}{2}x^2 + \cdots + (-1)^{n-1}\frac{1}{n}x^n$$

and the remainder is given by

$$R_n(x) = \frac{(-1)^n}{(1+c)^{n+1}(n+1)}x^{n+1}, \ c \in (0,x).$$

Note that if $n = 2k$, $R_{2k} > 0$ and if $n = 2k+1$, $R_{2k+1}(x) < 0$. Therefore, the inequality is established.

Theorem 4.3.3 *Let I be a neighborhood of the point a and suppose that the function $f : I \to \mathbb{R}$ has derivatives of all orders. Suppose also that there are positive numbers ρ and M such that the interval $(a-\rho, a+\rho)$ is contained in I and for every natural number n and every point x in $(a-\rho, a+\rho)$,*

$$|f^{(n)}(x)| \le M^n.$$

Then

$$f(x) = \sum_{n=0}^{\infty} \frac{f^{(n)}(a)}{n!}(x-a)^n, \quad \textit{for } |x-a| < \rho.$$

Example 4.3.10 Note that for any number c, $\lim_{n\to\infty} \dfrac{c^n}{n!} = 0$.

(1) $e^x = \sum_{n=0}^{\infty} \dfrac{x^k}{k!} \ \forall x \in \mathbb{R}$.

(2) $\ln(1+x) = \sum_{k=1}^{\infty} \dfrac{(-1)^{(k+1)}x^k}{k}$, if $-1 < x \le 1$.

Example 4.3.11 Using Taylor series, we can evaluate

$$\lim_{x\to 0} \frac{\sin x - \tan x}{x^3}.$$

Since

$$\sin x = x - \frac{x^3}{3!} + \frac{x^5}{5!} - \cdots, \quad \tan x = x + \frac{x^3}{3} + \frac{2x^5}{15} + \cdots,$$

we have

$$\sin x - \tan x = -\frac{x^3}{2} - \frac{x^5}{8} - \cdots.$$

Therefore,

$$\lim_{x\to 0} \frac{\sin x - \tan x}{x^3} = \lim_{x\to 0} \frac{x^3\left(-\frac{1}{2} - \frac{x^2}{8} - \cdots\right)}{x^3} = -\frac{1}{2}.$$

Remark 4.3.5 (Euler's Formula) For any real number x,

$$e^{ix} = \cos x + i \sin x.$$

using Taylor series of $\sin x$ and $\cos x$. Hence,

$$e^{i\pi} + 1 = 0.$$

We extend the Binomial theorem to the case of exponents that are not necessarily positive integers.

Theorem 4.3.4 (The Binomial Series) *Let m be any integer. Then,*

$$(1+x)^m = \sum_{k=0}^{\infty} \binom{m}{k} x^k, \quad \textit{for } -1 < x < 1,$$

where we define

$$\binom{m}{k} = \frac{m(m-1)(m-2)\cdots(m-k+1)}{k!}, \quad \textit{and } \binom{m}{0} = 1.$$

Remark 4.3.6 (1) In fact, the theorem holds for any real number m.

(2) When m is a negative integer, we call the Binomial series **negative Binomial expansion** , which is related to the negative Binomial random variable that can be found in probability and statisics.

(3) Theorem 4.3.4 can be shown by the Taylor expansion of $(1+x)^m$ about $x = 0$.

Example 4.3.12

(1) If $m = -1$, we have

$$\binom{-1}{k} = \frac{-1(-2)(-3)\cdots(-1-k+1)}{k!} = (-1)^k \frac{k!}{k!} = (-1)^k.$$

Thus, for $-1 < x < 1$,

$$(1+x)^{-1} = \sum_{k=0}^{\infty} \binom{-1}{k} x^k = \sum_{k=0}^{\infty} (-x)^k = 1 - x + x^2 - x^3 + \cdots + (-1)^k x^k + \cdots.$$

(2) If $m = -n$, $n \in \mathbb{N}$,

$$
\begin{aligned}
(1+x)^m &= \sum_{k=0}^{\infty} \binom{-n}{k} x^k \\
&= \sum_{k=0}^{\infty} (-1)^k \frac{n(n+1)\cdots(n+k-1)}{k!} x^k \\
&= \sum_{k=0}^{\infty} (-1)^k \binom{n+k-1}{k} x^k \\
&= \sum_{l=n}^{\infty} \binom{l-1}{n-1} (-x)^{l-n}.
\end{aligned}
$$

4.4 Suggested Exercises

1. Find the sum of the infinite series $\sum_{n=1}^{\infty} a_n$ if

   (a) $a_n = \dfrac{2}{n(n+1)} + \dfrac{1}{3^n}$.

   (b) $s_n = \sum_{k=1}^{n} a_k = \dfrac{n-1}{n+1}$.

2. Show that the power series $\sum_{n=2}^{\infty} \dfrac{\log n}{n^2} x^n$ converges if $|x| < 1$ and diverges if $|x| > 1$.

3. (Logarithmic p-series)

   (a) Show that $\int_2^{\infty} \dfrac{1}{x(\log x)^p} dx$, ($p$ a positive constant) converges if and only if $p > 1$.

   (b) What can you say about the convergence of the series $\sum_{n=2}^{\infty} \dfrac{1}{n(\log n)^p}$ based on the fact in (a)?

4. Show that if $\sum_{n=1}^{\infty} a_n$ converges, then $\sum_{n=1}^{\infty} \left(\dfrac{1+\sin(a_n)}{2}\right)^n$ converges.

5. Does the following series converge or not? Give reasons to your answer.

   (a) $\sum_{n=1}^{\infty} \dfrac{(n!)^2}{(2n)!}$.

   (b) $\sum_{n=1}^{\infty} \dfrac{1}{n\sqrt{n+1}}$.

   (c) $\sum_{n=1}^{\infty} \dfrac{1}{\sqrt{n(n+1}}$.

6. For what values of x does the following power series converge?

$$\sum_{n=1}^{\infty} (-1)^{n-1} \frac{x^n}{n}.$$

7. Investigate the convergence of $\sum_{n=1}^{\infty} \frac{2^n+5}{3^n+1}$.

8. Investigate the following series if it converges or diverges.

$$\sum_{n=1}^{\infty} \frac{(\log n)^2}{n^3}.$$

9. Determine if the following series is convergent or divergent.

   (a) $\sum_{n=1}^{\infty} \frac{n-2}{2n+3}$.

   (b) $\sum_{n=1}^{\infty} \frac{\sqrt{n^3+2}}{2n^3+3n^2+1}$.

   (c) $\sum_{n=1}^{\infty} ne^{-n^2}$.

   (d) $\sum_{n=1}^{\infty} (-1)^n \frac{n^3}{n^4+1}$.

   (e) $\sum_{n=1}^{\infty} \frac{3^k}{k!}$.

   (f) $\sum_{n=1}^{\infty} \frac{1}{3+4^n}$.

10. Find the radius of convergence and the interval of convergence of the series

   (a) $\sum_{n=0}^{\infty} \frac{(-4)^n x^n}{\sqrt{n+2}}$.

   (b) $\sum_{n=0}^{\infty} \frac{2n(x+2)^n}{5^{n+1}}$.

11. Does the following series converge or not? Give reasons to your answer.

   (a) $\sum_{n=1}^{\infty} \frac{n}{n^2+1}$. (b) $\sum_{n=1}^{\infty} \frac{n^{1/n}}{n^2}$.

12. Evaluate

$$\sum_{n=1}^{\infty} -\frac{(1/2)^n}{n}.$$

13. Evaluate

$$\lim_{x\to 0} \frac{e^x - 1 - x - \frac{1}{2!}x^2 - \cdots - \frac{1}{(n-1)!}x^{n-1}}{x^n} \quad \text{for } n = 3, 4, 5, \ldots,$$

without using L'Hospital's rule.

14. Suppose that $f(x) = \sum_{n=0}^{\infty} a_n x^n$ converges for all x in an open interval $(-c, c)$. Show that

(a) If f is even, then $a_1 = a_3 = a_5 = \cdots = 0$, i.e. the Taylor series for f at $x = 0$ contains only even powers of x.

(b) If f is odd, then $a_0 = a_2 = a_4 = \cdots = 0$, i.e. the Taylor series for f at $x = 0$ contains only odd powers of x.

(c) Find the Taylor polynomial of order 3 for the functions given below at 0.

$$f(x) = x \tan x, \quad f(x) = e^{\cos x}.$$

15. Evaluate the limit

(a) $\lim_{x\to 0} \dfrac{x - \tan^{-1}(x)}{2x^3}$.

(b) $\lim_{x\to 0} \dfrac{\sin x - x + \frac{1}{6}x^3}{2x^5}$.

16. Suppose that the function $f : \mathbb{R} \to \mathbb{R}$ has derivatives of all orders and that

$$\begin{cases} f'(x) + 2xf(x) = 0, & \text{for all } x, \\ f(0) = 1. \end{cases}$$

(a) Find a formula for the coefficients of the nth Taylor polynomial for f at $x = 0$.

(b) Show that the Taylor series expansion generated by f at $x = 0$ converges at every point in $\mathbb{R}$, $\forall x \in \mathbb{R}$.

(c) Using (a) and (b), deduce that

$$\int_0^1 f(x)dx = 1 - \frac{1}{3} + \frac{1}{10} - \frac{1}{42} + \cdots + \frac{(-1)^n}{(2n+1)n!} + \cdots.$$

(d) Does $\int_0^\infty f(x)dx$ converge? If it converges, then using a double integral in a polar form evaluate $\displaystyle\int_{-\infty}^{\infty} f(x)dx$.

(e) Evaluate $\displaystyle\int_0^\infty x^2 f(x)dx$.

17. Show that

(a)

$$e^x + e^{-x} = 2\sum_{n=0}^{\infty} \frac{x^{2n}}{(2n)!}, \quad \text{for } x \in \mathbb{R}.$$

(b)

$$\tan^{-1} x = x - \frac{x^3}{3} + \frac{x^5}{5} - \frac{x^7}{7} + \cdots, \quad \text{for } |x| < 1.$$

18. Use the Taylor series to evaluate the following limit:

$$\lim_{t\to 0} \frac{1 - \cos t - \frac{t^2}{2}}{t^4}.$$

19. Let $f(x)$ denote a function defined as an infinite series:

$$f(x) = \sum_{k=0}^{\infty} \binom{m}{k} x^k \quad \text{if } -1 < x < 1,$$

where m is an integer.

(a) Show that $f'(x) = \dfrac{mf(x)}{1+x}, \quad -1 < x < 1.$

(b) Define $g(x) = (1+x)^{-m} f(x)$ and show that $g'(x) = 0$.

(c) Conclude that $f(x) = (1+x)^m$.

20. Suppose that $|x| < 1$. Show that

(a) $\dfrac{1}{1+x^2} = \displaystyle\sum_{n=0}^{\infty} (-1)^n x^{2n}.$

(b) $\tan^{-1}(x) = \displaystyle\sum_{n=0}^{\infty} (-1)^n \frac{1}{2n+1} x^{2n+1}.$

Then, conclude that

$$\pi = 2\sqrt{3}\left(1 - \frac{1}{3\cdot 3} + \frac{1}{5\cdot 3^2} - \frac{1}{7\cdot 3^3} + \cdots\right).$$

21. Suppose that we would like to express $\log\frac{1+x}{1-x}$ as a power series. For this purpose, we consider the Taylor series expansion of $\log\frac{1+x}{1-x}$. However, instead of using the Taylor series of $\log\frac{1+x}{1-x}$ directly, we make use of the Taylor series expansions of $\log(1+x)$ and $\log(1-x)$ respectively.

    (a) Show that the following infinite series converges for $-1 < x < 1$.

    $$\sum_{n=1}^{\infty}(-1)^{n-1}\frac{x^n}{n}.$$

    You can consider either a suitable convergence test for infinite series or so-called 'term by term differentiation/integration'. Does it also converge when $x = 1$?

    (b) Show that the Taylor series expansion of $\log(1+x)$ is the same as the result in (a).

    (c) Show that the Taylor series expansion of

    $$\log\frac{1+x}{1-x} = 2\sum_{n=0}^{\infty}\frac{x^{2n+1}}{2n+1}, \quad |x| < 1.$$

22. Suppose that the function $f : \mathbb{R} \to \mathbb{R}$ has derivatives of all orders and that

    $$\begin{cases} f'(x) + xf(x) = 0, & \text{for all } x, \\ f(0) = 1. \end{cases}$$

    (a) Find a formula for the coefficients of the nth Taylor polynomial for f at $x = 0$.

    (b) Show that the Taylor expansion converges at every point in $\mathbb{R}$, $\forall x \in \mathbb{R}$.

    (c) Does $\int_0^\infty f'(x)dx$ converge? If it converges, then evaluate

    $$\int_{-\infty}^{\infty} f'(x)dx.$$

    (d) Evaluate

    $$\int_0^\infty x^2 f(x)dx.$$

Chapter 5

Partial Differentiation

In statistical inference, there emerges the "likelihood" which is a function of parameters (variables). For the case that the likelihood consists of two parameters (variables), we extend the derivative of single-variable function to the pair of "partial derivatives," that will be introduced in this chapter. Partial derivatives of the function of three or more variables will not be covered, even though the extension to the general case is not difficult.

5.1 Functions of Two Variables

A function f of two variables x and y is the rule of assigning z in $\mathbb{R}$ for each two-dimensional point (x, y) on the plane. That is,

$$f : (x, y) \to z, \quad \text{for } x, y, z \in \mathbb{R}.$$

More broadly, we may consider a function f of p variables $x_1, \cdots, x_p$ that assigns z in $\mathbb{R}$ for each p-tuple $(x_1, \cdots, x_p)$ in $\mathbb{R}^p$. That is,

$$f : (x_1, \cdots, x_p) \to z, \quad \text{for } x_1, \cdots, x_p, z \in \mathbb{R}.$$

Example 5.1.1 (Bivariate normal density function)

$$f(x, y) = \frac{1}{2\pi} e^{-\frac{x^2+y^2}{2}}, \quad -\infty < x < \infty, \quad -\infty < y < \infty. \tag{5.1.1}$$

This function is known as the **bivariate normal density** of two independent standard normal variables X and Y. At $(0, 0)$, f takes $\frac{1}{2\pi}$: $f(0, 0) = \frac{1}{2\pi}$. Furthermore, $f(1, 1) = \frac{1}{2\pi} e^{-1}$ and $f(\sqrt{2}, 0) = \frac{1}{2\pi} e^{-1} = f(0, \sqrt{2})$.

Remark 5.1.1 3D plot and contour plot provide useful visual displays for $z = f(x, y)$. For instance, Figure 5.1 is the 3D plot for $z = f(x, y)$ of Equation (5.1.1). We see a unimodal peak at the center.

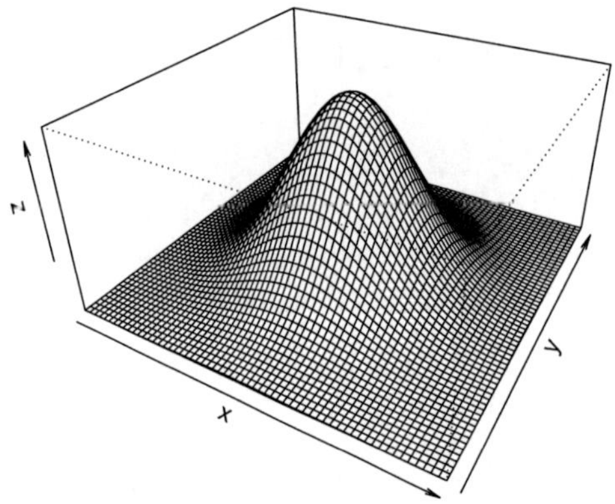

Figure 5.1: 3D plot of $f(x, y) = \frac{1}{2\pi} e^{-\frac{x^2+y^2}{2}}$

Definition 5.1.1 *A* **contour line** *(curve) K_c is defined as the set of all (x, y)'s at which f takes the same value, say c: $K_c = \{(x, y) : f(x, y) = c\}$. Contour plot consists of contour lines (curves). Figure 5.2 shows contour curves of $z = f(x, y)$ of Equation (5.1.1).*

Definition 5.1.2 *A function $f(x, y)$ approaches the limit L as (x, y) approaches (x_0, y_0) if, for every $\epsilon > 0$, there exists $\delta > 0$ such that*

$$0 < \sqrt{(x - x_0)^2 + (y - y_0)^2} < \delta \quad \Rightarrow \quad |f(x, y) - L| < \epsilon.$$

Remark 5.1.2 In such a case, we write $\lim_{(x,y)\to(x_0,y_0)} f(x, y) = L$.

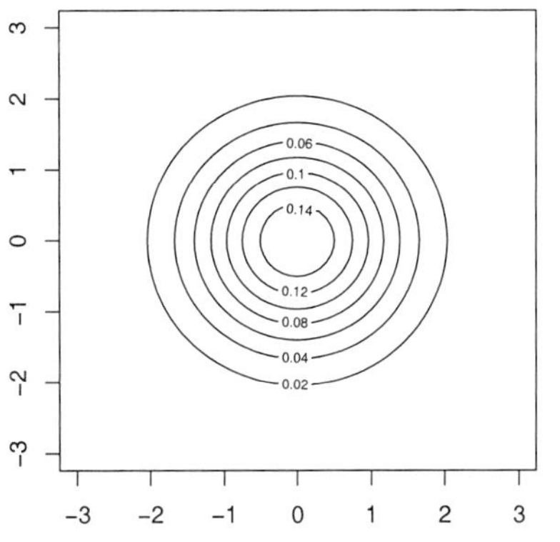

Figure 5.2: Contour plot of $f(x,y) = \frac{1}{2\pi}e^{-\frac{x^2+y^2}{2}}$

Example 5.1.2 (Continued from Example 5.1.1) For $f(x,y) = \frac{1}{2\pi}e^{-\frac{x^2+y^2}{2}}$,

$$\lim_{(x,y)\to(0,0)} f(x,y) = \frac{1}{2\pi}$$

since, by taking $\delta = \sqrt{\epsilon}$, $0 < \sqrt{x^2+y^2} < \delta$ implies that

$$\left|\frac{1}{2\pi}e^{-\frac{x^2+y^2}{2}} - \frac{1}{2\pi}\right| = \frac{1}{2\pi}(1 - e^{-\frac{x^2+y^2}{2}}) < \frac{1}{2\pi}\frac{x^2+y^2}{2} < \delta^2 = \epsilon.$$

In above, we used the inequality $e^{-u} > 1-u$, for $u > 0$.

Example 5.1.3

$$\lim_{(x,y)\to(0,0)} \frac{x^4+y^4}{x^2+y^2} = 0.$$

Write x and y in polar coordinates: $x = r\cos\theta,\ y = r\sin\theta$. Then "$(x,y) \to (0,0)$" is equivalent to "$r \to 0$". Therefore

$$\lim_{(x,y)\to(0,0)} \frac{x^4+y^4}{x^2+y^2} = \lim_{r\to 0}\frac{r^4(\cos^4\theta + \sin^4\theta)}{r^2} = 0.$$

Definition 5.1.3 *A function $f(x, y)$ is continuous at (x_0, y_0), if*

1) f is defined at (x_0, y_0),

2) $\lim_{(x,y)\to(x_0,y_0)} f(x, y)$ exists,

3) $\lim_{(x,y)\to(x_0,y_0)} f(x, y) = f(x_0, y_0)$.

A function is continuous if it is continuous everywhere (on its domain).

Example 5.1.4 (Continued from Example 5.1.3) Let

$$f(x,y) = \frac{x^4 + y^4}{x^2 + y^2}, \quad \text{if} \quad (x, y) \neq (0, 0).$$

Since $\lim_{(x,y)\to(0,0)} f(x, y) = 0$, f is continuous everywhere if we define f at $(0, 0)$ to be zero.

Example 5.1.5 Let

$$f(x,y) = \frac{2x\,y}{x^2 + y^2}, \quad \text{if} \quad (x, y) \neq (0, 0).$$

We consider two paths along which (x, y) approaches $(0, 0)$.

1) $x = y$ or $x = t, y = t$. Along this path,

$$\lim_{(x,y)\to(0,0)} \frac{2x\,y}{x^2 + y^2} = \lim_{t\to 0} \frac{2t^2}{t^2 + t^2} = 1.$$

2) $x = -y$ or $x = t, y = -t$. Along this path,

$$\lim_{(x,y)\to(0,0)} \frac{2x\,y}{x^2 + y^2} = \lim_{t\to 0} \frac{-2t^2}{t^2 + t^2} = -1.$$

Hence $\lim_{(x,y)\to(0,0)} f(x, y)$ does not exist. See Figure 5.3 for the 3D plot of $z = f(x, y)$.

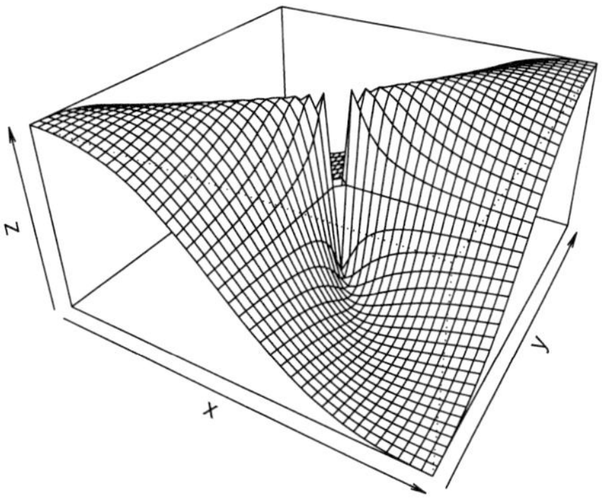

Figure 5.3: 3D plot of $f(x, y) = \frac{2xy}{x^2+y^2}$

5.2 Partial Derivatives

Definition 5.2.1 *For a continuous function $f(x, y)$, local characteristics at (x_0, y_0) can be captured by partial derivatives, $\frac{\partial f}{\partial x}|_{(x_0,y_0)}$ and $\frac{\partial f}{\partial y}|_{(x_0,y_0)}$, defined as follows.*

If the limits exist,

$$\frac{\partial f}{\partial x}|_{(x_0,y_0)} = \lim_{(x,y)\to(x_0,y_0)} \frac{f(x_0+h_1, y_0) - f(x_0, y_0)}{h_1},$$

$$\frac{\partial f}{\partial y}|_{(x_0,y_0)} = \lim_{(x,y)\to(x_0,y_0)} \frac{f(x_0, y_0+h_2) - f(x_0, y_0)}{h_2}.$$

Remark 5.2.1 (1) The partial derivative $\frac{\partial f}{\partial x}|_{(x_0,y_0)}$ of $f(x, y)$ with respect to x is the slope of the tangent line to $f(x, y_0)$, a function of single variable x, at $x = x_0$.

(2) Similarly, the partial derivative $\frac{\partial f}{\partial y}|_{(x_0,y_0)}$ of $f(x, y)$ with respect to y is the slope of the tangent line to $f(x_0, y)$, a function of single variable y, at $y = y_0$.

(3) The tangent plane of $z = f(x, y)$ at (x_0, y_0) is given by

$$z = f(x_0, y_0) + \frac{\partial f}{\partial x}|_{(x_0,y_0)}(x - x_0) + \frac{\partial f}{\partial y}|_{(x_0,y_0)}(y - y_0).$$

(4) In Figure 5.4, the tangent plane is represented as a panel touching $z = f(x, y)$ at O: (x_0, y_0), with the sides OP and OQ, of which the slopes are $\frac{\partial f}{\partial x}|_{(x_0,y_0)}$ and $\frac{\partial f}{\partial y}|_{(x_0,y_0)}$ with respect to x-axis and y-axis.

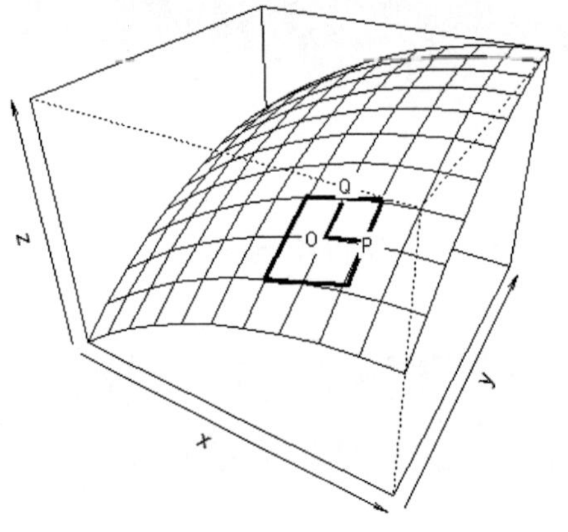

Figure 5.4: Tangent plane of $z = f(x, y)$ at O: (x_0, y_0).

Example 5.2.1 Let $z = f(x, y) = x^2 + y^2$. Then $\frac{\partial f}{\partial x} = 2x$. $\frac{\partial f}{\partial y} = 2y$. Hence at the point O: $(1, 2)$, $\frac{\partial f}{\partial x}|_{(1,2)} = 2$ and $\frac{\partial f}{\partial y}|_{(1,2)} = 4$. Consequently, the tangent plane at O: $(1, 2)$ is

$$z = 5 + 2(x - 1) + 4(y - 2) = 2x + 4y - 5.$$

Remark 5.2.2 (1) $\dfrac{\partial f}{\partial x}$ is a shorthand of the partial derivative with respect to x evaluated at an arbitrary point (x, y). Similarly, $\dfrac{\partial f}{\partial y}$ denotes the partial derivative with respect to y evaluated at an arbitrary point (x, y).

(2) There are several alternative notations of the partial derivatives:

(a) For $\dfrac{\partial f}{\partial x}$, one may use $D_x(f)$, $D_1(f)$ or $f_x(x, y)$.

(b) For $\dfrac{\partial f}{\partial y}$, one may use $D_y(f)$, $D_2(f)$ or $f_y(x, y)$.

Definition 5.2.2 Second-order partial derivatives $\frac{\partial^2 f}{\partial x^2}, \frac{\partial^2 f}{\partial y\,\partial x}, \frac{\partial^2 f}{\partial x\,\partial y}, \frac{\partial^2 f}{\partial y^2}$ *of* f *are defined as follows:*

$$\frac{\partial^2 f}{\partial x^2} = \frac{\partial}{\partial x}\left(\frac{\partial f}{\partial x}\right), \quad \frac{\partial^2 f}{\partial y\,\partial x} = \frac{\partial}{\partial y}\left(\frac{\partial f}{\partial x}\right),$$

$$\frac{\partial^2 f}{\partial x\,\partial y} = \frac{\partial}{\partial x}\left(\frac{\partial f}{\partial y}\right), \quad \frac{\partial^2 f}{\partial y^2} = \frac{\partial}{\partial y}\left(\frac{\partial f}{\partial y}\right).$$

(1) $\frac{\partial^2 f}{\partial x^2}$ *is the partial derivative of* $\frac{\partial f}{\partial x}$ *with respect to* x.

(2) $\frac{\partial^2 f}{\partial y\,\partial x}$ *is the partial derivative of* $\frac{\partial f}{\partial x}$ *with respect to* y.

(3) $\frac{\partial^2 f}{\partial x\,\partial y}$ *is the partial derivative of* $\frac{\partial f}{\partial y}$ *with respect to* x.

(4) $\frac{\partial^2 f}{\partial y^2}$ *is the partial derivative of* $\frac{\partial f}{\partial y}$ *with respect to* y.

Example 5.2.2 For $f(x, y) = 2x^2 + y^2 + x\,y$,

$$\frac{\partial f}{\partial x} = 4x + y \quad\Rightarrow\quad \frac{\partial^2 f}{\partial x^2} = \frac{\partial}{\partial x}\left(\frac{\partial f}{\partial x}\right) = 4, \quad \frac{\partial^2 f}{\partial y\,\partial x} = \frac{\partial}{\partial y}\left(\frac{\partial f}{\partial x}\right) = 1,$$

$$\frac{\partial f}{\partial y} = 2y + x \quad\Rightarrow\quad \frac{\partial^2 f}{\partial x\,\partial y} = \frac{\partial}{\partial x}\left(\frac{\partial f}{\partial y}\right) = 1, \quad \frac{\partial^2 f}{\partial y^2} = \frac{\partial}{\partial y}\left(\frac{\partial f}{\partial y}\right) = 2.$$

In this example, we may note that

$$\frac{\partial^2 f}{\partial y\,\partial x} = \frac{\partial^2 f}{\partial x\,\partial y},$$

which is generally true provided that $\frac{\partial^2 f}{\partial y\,\partial x}$ and $\frac{\partial^2 f}{\partial x\,\partial y}$ can be defined and continuous.

Definition 5.2.3 *A function* $z = f(x, y)$ *is* **differentiable** *at* (x_0, y_0), *if* $f_x(x_0, y_0)$ *and* $f_y(x_0, y_0)$ *exist and* Δz *can be expressed as*

$$\Delta z = f_x(x_0, y_0)\,\Delta x + f_y(x_0, y_0)\,\Delta y + \epsilon_1 \Delta x + \epsilon_2 \Delta y,$$

where ϵ_1 *and* ϵ_2 *go to 0, as* Δx *and* Δy *go to 0.*

The following theorem provides a sufficient condition that guarantees the differentiability of f.

Theorem 5.2.1 *If the partial derivatives $f_x(x,y)$ and $f_y(x,y)$ of $f(x,y)$ are continuous, then f is differentiable at every point.*

Example 5.2.3 (Continued from Example 5.2.2) For $f(x,y) = 2x^2 + y^2 + x\,y$,

$$f_x(x,y) = 4x + y, \quad f_y(x,y) = 2y + x.$$

We may note that $4x+y$ and $2y+x$ are continuous. Hence, by Theorem 5.2.1, $f(x,y) = 2x^2 + y^2 + x\,y$ is differentiable.

Example 5.2.4 For $f(x,y) = e^{-\frac{1}{2}(x^2+y^2)}$,

$$f_x(x,y) = -xe^{-\frac{1}{2}(x^2+y^2)}, \quad f_y(x,y) = -ye^{-\frac{1}{2}(x^2+y^2)}.$$

We may note that f_x and f_y are continuous. Hence, by Theorem 5.2.1, $f(x,y)$ is differentiable.

Suppose that x and y are differentiable functions of t. Write $x = x(t)$ and $y = y(t)$.

Theorem 5.2.2 (The Chain Rule) *If the partial derivatives $f_x(x,y)$ and $f_y(x,y)$ of $f(x,y)$ are continuous, then $z = f(x(t), y(t))$ is differentiable with respect to t and*

$$z'(t) = f_x(x(t), y(t))x'(t) + f_y(x(t), y(t))y'(t)$$

or

$$\frac{dz}{dt} = \frac{\partial f}{\partial x}\frac{dx}{dt} + \frac{\partial f}{\partial y}\frac{dy}{dt}.$$

Proof Since $z = f(x,y)$ has continuous partial derivatives f_x and f_y,

$$\Delta z = f_x(x_0, y_0)\,\Delta x + f_y(x_0, y_0)\,\Delta y + \epsilon_1 \Delta x + \epsilon_2 \Delta y,$$

where ϵ_1 and ϵ_2 go to 0, as Δx and Δy go to 0. Hence

$$\frac{\Delta z}{\Delta t} = f_x(x_0, y_0)\frac{\Delta x}{\Delta t} + f_y(x_0, y_0)\frac{\Delta y}{\Delta t} + \epsilon_1\frac{\Delta x}{\Delta t} + \epsilon_2\frac{\Delta y}{\Delta t}.$$

As $\Delta t \to 0$, $\Delta x \to 0$ and $\Delta y \to 0$. Hence, we have

$$\frac{dz}{dt} = f_x\frac{dx}{dt} + f_y\frac{dy}{dt} + 0\cdot\frac{dx}{dt} + 0\cdot\frac{dy}{dt}.$$

Example 5.2.5 Let $f(x, y) = x\,y,\ x = \cos t,\ y = \sin t$. Then, for $z = f(x(t), y(t))$,

$$\begin{aligned}\frac{dz}{dt} = \frac{\partial f}{\partial x}\frac{dx}{dt} + \frac{\partial f}{\partial y}\frac{dy}{dt} &= y\frac{dx}{dt} + x\frac{dy}{dt} = \sin t\,(-\sin t) + \cos t\,(\cos t) \\ &= \cos^2 t - \sin^2 t.\end{aligned}$$

The derivative $\frac{dz}{dt}$ at $t = \frac{\pi}{4}$ is

$$\cos^2\frac{\pi}{4} - \sin^2\frac{\pi}{4}$$

which is 0.

Now, consider the situation in which x and y are differentiable functions of s and t. Write $x = x(s, t)$ and $y = y(s, t)$.

Theorem 5.2.3 (The Chain Rule, An Extension) *If the partial derivatives f_x and f_y of $f(x, y)$ are continuous, then $z = f(x(s, t), y(s, t))$ is differentiable with respect to s and t. Furthermore,*

$$\frac{\partial z}{\partial s} = \frac{\partial f}{\partial x}\frac{\partial x}{\partial s} + \frac{\partial f}{\partial y}\frac{\partial y}{\partial s}, \quad \frac{\partial z}{\partial t} = \frac{\partial f}{\partial x}\frac{\partial x}{\partial t} + \frac{\partial f}{\partial y}\frac{\partial y}{\partial t}.$$

Proof Consider a special case in which t is a fixed constant. Then the first equation is obtained as a result of Theorem 5.2.2. Similarly, the second equation can be obtained.

Example 5.2.6 Let $f(x, y) = x\,y,\ x = \cos(s + t),\ y = \sin(s - t)$.
For $z = f(x(s, t), y(s, t))$,

$$\begin{aligned}\frac{\partial z}{\partial s} &= \frac{\partial f}{\partial x}\frac{\partial x}{\partial s} + \frac{\partial f}{\partial y}\frac{\partial y}{\partial s} = y\frac{\partial x}{\partial s} + x\frac{\partial y}{\partial s} \\ &= \sin(s - t)(-\sin(s + t)) + \cos(s + t)\cos(s - t) \\ &= \cos(s + t)\cos(s - t) - \sin(s + t)\sin(s - t) = \cos(2s),\end{aligned}$$

$$\begin{aligned}\frac{\partial z}{\partial t} &= \frac{\partial f}{\partial x}\frac{\partial x}{\partial t} + \frac{\partial f}{\partial y}\frac{\partial y}{\partial t} = y\frac{\partial x}{\partial t} + x\frac{\partial y}{\partial t} \\ &= \sin(s - t)(-\sin(s + t)) - \cos(s + t)\cos(s - t) \\ &= -\cos(s + t)\cos(s - t) - \sin(s + t)\sin(s - t) = -\cos(2t).\end{aligned}$$

Suppose that the relationship of y to x is defined implicitly via some functional relationship $F(x, y) = 0$. The following theorem states that $\frac{dy}{dx}$ can be expressed with the partial derivatives of F with respect to x and y.

Theorem 5.2.4 *Suppose that $F(x, y)$ is differentiable and that $F(x, y) = 0$ defines y as a differentiable function of x. Then, at any point where $F_y \neq 0$,*

$$\frac{dy}{dx} = -\frac{F_x}{F_y}.$$

Proof Let $z = F(x, y)$. Since y is a function of x (via $F(x, y) = 0$), z is a function of x. Thus

$$\frac{dz}{dx} = F_x \frac{dx}{dx} + F_y \frac{dy}{dx} \quad \text{or} \quad 0 = F_x \cdot 1 + F_y \frac{dy}{dx}.$$

$$\therefore \quad \frac{dy}{dx} = -\frac{F_x}{F_y} \quad \text{if } F_y \neq 0.$$

Example 5.2.7 Let $x^2 + y^2 = 1$. By differentiating both sides with respect to x,

$$2x + 2y\frac{dy}{dx} = 0.$$

$$\therefore \quad \frac{dy}{dx} = -\frac{x}{y} \quad \text{if } y \neq 0.$$

Hence, the tangent line of the circle $x^2 + y^2 = 1$ at (x_0, y_0) is

$$y - y_0 = -\frac{x_0}{y_0}(x - x_0), \quad \text{or} \quad x_0(x - x_0) + y_0(y - y_0) = 0.$$

$$\therefore \quad x_0\,x + y_0\,y = x_0^2 + y_0^2 = 1.$$

For a function $f(x, y)$ of two variables x and y, there are two directions of the change, one along the x-axis and another along the y-axis. Now, we define the composite measure of the change.

Definition 5.2.4 *The change rate of $f(x, y)$ along the unit vector $\mathbf{u}$ is called the* **directional derivative** *of f in the direction of $\mathbf{u} = (u_1, u_2), u_1^2 + u_2^2 = 1$. The directional derivative at (x_0, y_0) is*

$$\frac{df}{dt} = \lim_{t \to 0} \frac{1}{t}\{f(x_0 + tu_1, y_0 + tu_2) - f(x_0, y_0)\}$$

provided that the limit exists. By the chain rule,

$$\frac{df}{dt} = \frac{\partial f}{\partial x}u_1 + \frac{\partial f}{\partial y}u_2.$$

Definition 5.2.5 *The* **gradient** $\bigtriangledown f$ *of* $f(x,y)$ *at* (x_0, y_0) *is the vector of directional derivatives in the direction of* x*-axis and* y*-axis:*

$$\bigtriangledown f = \frac{\partial f}{\partial x}\mathbf{i} + \frac{\partial f}{\partial y}\mathbf{j},$$

where $\mathbf{i} = (1,0)$, $\mathbf{j} = (0,1)$. *Thus,*

$$\bigtriangledown f = \left(\frac{\partial f}{\partial x}, \frac{\partial f}{\partial y}\right).$$

Example 5.2.8 (Continued from Example 5.2.7) Let $f(x,y) = x^2 + y^2$. Then

$$\bigtriangledown f = \left(\frac{\partial f}{\partial x}, \frac{\partial f}{\partial y}\right) = (2x, 2y).$$

$$\therefore \quad \bigtriangledown f(1,1) = (2,2),\ \bigtriangledown f(1,0) = (2,0),\ \bigtriangledown f(0,1) = (0,2).$$

Remark 5.2.3 Since the gradient $\bigtriangledown f$ is a vector, we can define the **inner product** of $\bigtriangledown f$ and the unit vector $\mathbf{u}\,(= (u_1, u_2), u_1^2 + u_2^2 = 1)$. We see that

$$\bigtriangledown f \cdot \mathbf{u} = \frac{\partial f}{\partial x} u_1 + \frac{\partial f}{\partial y} u_2,$$

which is identical to the directional derivative of f in the direction of $\mathbf{u}$. Hence, the direction vector in which the maximum gain occurs is given by

$$\mathbf{u}^+ = \frac{\bigtriangledown f}{\| \bigtriangledown f \|}.$$

Similarly, the direction vector in which the maximum loss occurs is given by

$$\mathbf{u}^- = -\frac{\bigtriangledown f}{\| \bigtriangledown f \|}.$$

Example 5.2.9 Let $f(x,y) = x\,y$. Then

$$\bigtriangledown f(1,1) = (1,1),\ \bigtriangledown f(\frac{3}{2}, \frac{2}{3}) = (\frac{2}{3}, \frac{3}{2}).$$

Therefore, the steepest ascents of f at P: $(1,1)$ and at Q: $(\frac{3}{2}, \frac{2}{3})$ are along the paths directed by the unit vector $(1,1)/\sqrt{2}$ and $(\frac{2}{3}, \frac{3}{2})/\sqrt{(\frac{2}{3})^2 + (\frac{3}{2})^2}$, respectively. See Figure 5.5.

Similarly, the steepest descents of f at P: $(1,1)$ and at Q: $(\frac{3}{2}, \frac{2}{3})$ are along the paths directed by the unit vector $-(1,1)/\sqrt{2}$ and $-(\frac{2}{3}, \frac{3}{2})/\sqrt{(\frac{2}{3})^2 + (\frac{3}{2})^2}$, respectively.

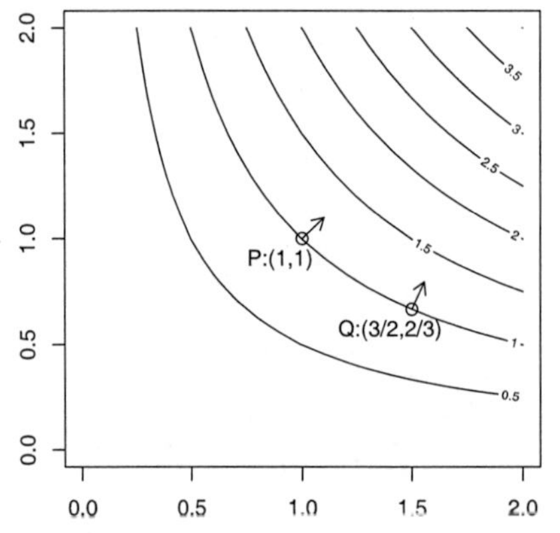

Figure 5.5: Contour curves of $f(x,y) = x\,y$ and gradient vectors at P: $(1,1)$ and at Q: $(\frac{3}{2},\frac{2}{3})$

5.3 Local Maxima and Minima of the Functions of Two Variables

The goal of this section is to find the maximum and/or the minimum of the function $f(x,y)$.

Definition 5.3.1 *Let $f(x,y)$ be a real-valued function defined on some domain D containing the point (x_0,y_0). Then*

1) $f(x_0,y_0)$ is a **local maximum** *of f if $f(x_0,y_0) \geq f(x,y)$ for all points (x,y) in an open circle centered at (x_0,y_0).*

2) $f(x_0,y_0)$ is a **local minimum** *of f if $f(x_0,y_0) \leq f(x,y)$ for all points (x,y) in an open circle centered at (x_0,y_0).*

Figure 5.6 is an illustration of local maximums of the function of two variables. The plot shows two peaks, that are local maximums, one of which is the global maximum.

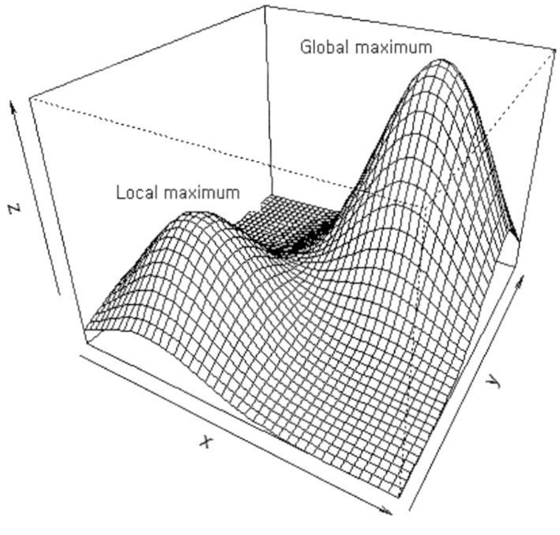

Figure 5.6: Local maximum and global maximum

Definition 5.3.2 *An* **interior point** (x_0, y_0) *at which a function* $f(x, y)$ *has zero partial derivatives is called a critical point of* f.

Definition 5.3.3 *A* **critical point** (x_0, y_0) *of* f *is called a saddle point, if every open circle centered at* (x_0, y_0) *contains a point* (x_1, y_1) *such that* $f(x_1, y_1) > f(x_0, y_0)$ *and a point* (x_2, y_2) *such that* $f(x_2, y_2) < f(x_0, y_0)$.

Example 5.3.1 Let $f(x, y) = x^2 - y^2$. Then P: $(0, 0)$ is the unique critical point of f and $f(0, 0) = 0$. Consider an open circle of radius $\delta > 0$: $x^2 + y^2 < \delta^2$. Since

$$f\left(\frac{\delta}{2}, 0\right) = \frac{\delta^2}{4} > f(0,0) \quad \text{and} \quad f\left(0, \frac{\delta}{2}\right) = -\frac{\delta^2}{4} < f(0,0),$$

P is a **saddle point**. See Figure 5.7.

Theorem 5.3.1 (Second Derivative Test for Local Maximum or Minimum) *Suppose that, for continuous* $f(x, y)$*, the first and second partial derivatives are continuous at* (x_0, y_0) *and that* $f_x(x_0, y_0) = f_y(x_0, y_0) = 0$. *Denote* $D = f_{xx} f_{yy} - f_{xy}^2$. *Then*

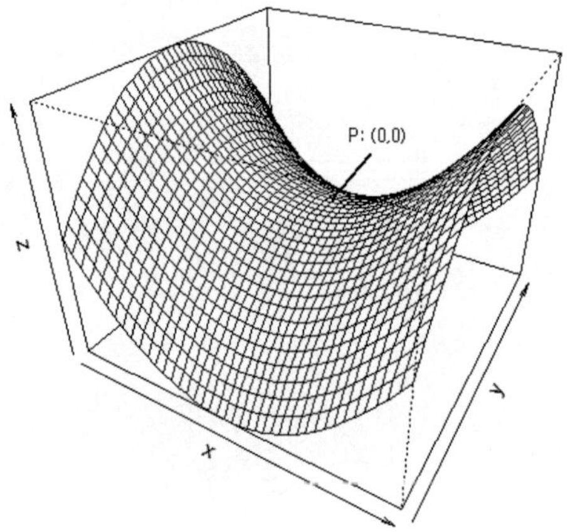

Figure 5.7: A saddle point P: $(0, 0)$ of $f(x, y) = x^2 - y^2$

1) if $f_{xx} < 0$ and $D > 0$, (x_0, y_0) is a local maximum of f.

2) if $f_{xx} > 0$ and $D > 0$, (x_0, y_0) is a local minimum of f.

3) if $D < 0$, (x_0, y_0) is a saddle point of f.

Remark 5.3.1 Theorem 5.3.1 is understandable, since it extends the case of single-variable functions, except the appearance of D. When the signs f_{xx} and f_{yy} are opposite, D is negative. We will not prove Theorem 5.3.1.

Example 5.3.2 (Continued from Example 5.3.1) Let $f(x, y) = x^2 - y^2$. Then $f_x(x, y) = 2x$ and $f_y(x, y) = -2y$. Hence, P: $(0, 0)$ is the unique critical point. At P: $(0, 0)$, $D = 2 \cdot (-2) < 0$. Therefore, P: $(0, 0)$ is a saddle point.

Example 5.3.3 Let $f(x, y) = x^2 + y^2 - c\,x\,y$, where c is some constant not equal to ± 2. Then $f_x(x, y) = 2x - cy$ and $f_y(x, y) = 2y - c\,x$. Hence, P: $(0, 0)$ is the unique critical point.

Now, we compute D. $f_{xx}(x, y) = 2$, $f_{yy}(x, y) = 2$, $f_{xy}(x, y) = -c$. $\therefore D = 4 - c^2$. Therefore, if $-2 < c < 2$, P: $(0, 0)$ is the local minimum. If $c < -2$ or $c > 2$, P: $(0, 0)$ is a saddle point. See Figure 5.8 for the contour curves of f for the case of $c = 4$.

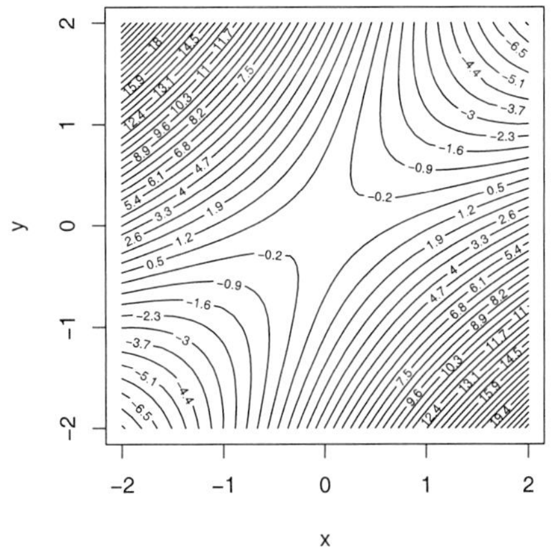

Figure 5.8: Level curves of $f(x, y) = x^2 + y^2 - c\,x\,y$, for $c = 4$

Example 5.3.4 Let $f(x, y) = x^2 + y^2 - x\,y - 2x$. Then $f_x(x, y) = 2x - y - 2$ and $f_y(x, y) = 2y - x$. Hence $(\frac{4}{3}, \frac{2}{3})$ is a critical point. Note that $f_{xx}(x, y) = 2$, $f_{yy}(x, y) = 2$, $f_{xy}(x, y) = -1$ and that $D = 3$. Since $f_{xx} > 0$ and $D > 0$, f has a local minimum at P: $(\frac{4}{3}, \frac{2}{3})$. $f(\frac{4}{3}, \frac{2}{3}) = -\frac{4}{3}$.

Example 5.3.5 (Continued from Example 5.3.4) Let $f(x, y) = x^2 + y^2 - x\,y - 2x$ on the triangle area A: $0 \le x, y \le 6$, $0 \le x + y \le 6$. Find the maximum and minimum of f on A.

Since P: $(\frac{4}{3}, \frac{2}{3})$ is the unique local minimum in the domain A, f has the minimum at P and $f(\frac{4}{3}, \frac{2}{3}) = -\frac{4}{3}$. The maximum could occur at the boundaries of A. On the line segment connecting O: $(0, 0)$ and Q: $(6, 0)$, the maximum occurs at Q: $(6, 0)$ and $f(6, 0) = 24$. On the line segment connecting O: $(0, 0)$ and R: $(0, 6)$, the maximum occurs at R: $(0, 6)$ and $f(0, 6) = 36$. Therefore, f has the maximum at R: $(0, 6)$ and $f(0, 6) = 36$. See Figure 5.9.

The second derivative test in Theorem 5.3.1 can be understood as the extension of the second derivative test for one-dimensional function by introducing the concept of the Hessian and the Hessian form in the following way. First, we define the quadratic form of a symmetric matrix $\boldsymbol{A}$ of order 2×2 for simplicity.

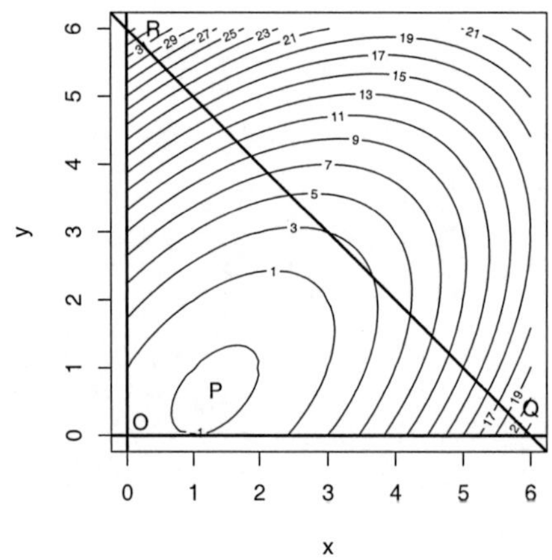

Figure 5.9: Level curves of $f(x, y) = x^2 + y^2 - x\,y - 2x$, $0 \le x, y \le 6$, $0 \le x + y \le 6$.

Definition 5.3.4 *Let* $\boldsymbol{A} = \begin{pmatrix} a & b \\ b & c \end{pmatrix}$ *be a symmetric matrix of order* 2×2*. Then, the function* $q(x, y) = (x, y)\boldsymbol{A}(x, y)^T$ *is called the a* **quadratic form** *in* (x, y),

$$q(x, y) = ax^2 + 2bxy + cy^2.$$

A quadratic form is said to be

(1) **positive definite** *if* $q(x, y) > 0$ *for all* $(x, y) \in \mathbb{R}^2$*. Equivalently, it is positive definite if* $a > 0$ *and* $\det(\boldsymbol{A}) = ac - b^2 > 0$*. We call matrix* $\boldsymbol{A}$ **positive definite matrix** *and denote* $\boldsymbol{A} > \boldsymbol{O}$*.*

(2) **negative definite** *if* $q(x, y) < 0$ *for all* $(x, y) \in \mathbb{R}^2$*. Equivalently, it is negative definite if* $a < 0$ *and* $\det(\boldsymbol{A}) = ac - b^2 > 0$*. We call matrix* $\boldsymbol{A}$ **negative definite matrix** *and denote* $\boldsymbol{A} < \boldsymbol{O}$*.*

Theorem 5.3.2 *Let* $\boldsymbol{A} = \begin{pmatrix} a & b \\ b & c \end{pmatrix}$ *be a symmetric matrix of order* 2×2*, and consider the quadratic form* $q(x, y) = ax^2 + 2bxy + cy^2$ *based on* $\boldsymbol{A}$*. Then,*

(1) If $\boldsymbol{A} > \boldsymbol{O}$, $(0,0)$ *is a global minimum of* $q(x,y)$.

(2) If $\boldsymbol{A} < \boldsymbol{O}$, $(0,0)$ *is a global maximum of* $q(x,y)$.

(3) If $\det(\boldsymbol{A}) < 0$, $(0,0)$ *is a saddle point of* $q(x,y)$.

Definition 5.3.5 *Suppose that the function* $z = f(x,y)$ *has continuous second derivatives on a neighborhood of a critical point* $(x,y) = (a,b)$. *The* **Hessian** *of* f *at* (a,b) *is the symmetric matrix of second derivatives*

$$\boldsymbol{H}_{(a,b)} = \begin{pmatrix} f_{xx}(a,b) & f_{xy}(a,b) \\ f_{yx}(a,b) & f_{yy}(a,b) \end{pmatrix}.$$

Remark 5.3.2 (1) Continuity of the second derivatives guarantees that $\boldsymbol{H}_{(a,b)}$ is symmetric.

(2) The **Hessian form** of f at (a,b) is the quadratic form associated with the Hessian. We often use the symbol $H_{(a,b)}$ for the Hessian form as well as the Hessian interchangeably. Thus,

$$H_{(a,b)}(x,y) = (x,y)\boldsymbol{H}_{a,b}(x,y)^T = f_{xx}(a,b)x^2 + 2f_{xy}(a,b)xy + f_{yy}(a,b)y^2.$$

(3) We can express the second derivative test in Theorem 5.3.1 in terms of the Hessian matrix $H_{(a,b)}$ and its determinant as in Theorem 5.3.3

Theorem 5.3.3 (Second Derivative Test for Local Maximum or Minimum) *Suppose that, for continuous* $f(x,y)$, *the first and second partial derivatives are continuous at* (x_0,y_0) *and that* $f_x(x_0,y_0) = f_y(x_0,y_0) = 0$. *Then*

1) if $\boldsymbol{H}_{x_0,y_0} < \boldsymbol{O}$, *then* (x_0,y_0) *is a local maximum of* f.

2) if $\boldsymbol{H}_{x_0,y_0} > \boldsymbol{O}$, *then* (x_0,y_0) *is a local minimum of* f.

3) if $\det(\boldsymbol{H}_{x_0,y_0}) < 0$, *then* (x_0,y_0) *is a saddle point of* f.

Example 5.3.6 For functions $f(x,y) = x^4 + y^4 - 4xy + 2$, the critical points are $(0,0)$, $(1,1)$, $(-1,-1)$ since the gradient of f is $\bigtriangledown f = (4x^3 - 4y, -4x + 4y^3)$. The Hessian matrix is given by

$$\boldsymbol{H}_{x,y} = \begin{pmatrix} 12x^2 & -4 \\ -4 & 12y^2 \end{pmatrix}.$$

Since $\boldsymbol{H}_{1,1} = \boldsymbol{H}_{-1,-1} > \boldsymbol{O}$ and $\det(\boldsymbol{H}_{0,0}) < 0$, $(1,1)$ and $(-1,-1)$ are local minimum while $(0,0)$ is a saddle point of f.

5.4 The Method of Lagrange Multipliers

Suppose that $f(x,y)$ and $g(x,y)$ are differentiable and $\bigtriangledown g \neq 0$ when $g(x,y)=0$. To find the local maximum (or minimum) values of $f(x,y)$ subject to the constraint $g(x,y)=0$ (if these exist), we need to determine the values of x, y and λ that simultaneously satisfy the equations

$$\bigtriangledown f = \lambda \bigtriangledown g, \quad g(x,y)=0. \tag{5.4.2}$$

Remark 5.4.1 Express $g(x,y)=0$ by parametric equations: $x=x(t), y=y(t)$. Then, on the curve $g(x,y)=0$, the function to be maximized (or minimized) is $f(x(t),y(t))$. Thus, at the critical points of f subject to $g(x,y)=0$,

$$\frac{df}{dt}=\frac{\partial f}{\partial x}\frac{dx}{dt}+\frac{\partial f}{\partial y}\frac{dy}{dt}\ (=\bigtriangledown f\cdot\mathbf{v})=0, \tag{5.4.3}$$

where $\mathbf{v}=\frac{dx}{dt}\,\mathbf{i}+\frac{dy}{dt}\,\mathbf{j}$. Since $g(x(t),y(t))=0$,

$$\frac{dg}{dt}=\frac{\partial g}{\partial x}\frac{dx}{dt}+\frac{\partial g}{\partial y}\frac{dy}{dt}\ (=\bigtriangledown g\cdot\mathbf{v})=0. \tag{5.4.4}$$

From Equations (5.4.3) and (5.4.4), $\bigtriangledown f$ is proportional to $\bigtriangledown g$: That is, $\bigtriangledown f = \lambda \bigtriangledown g$ for some non-zero constant λ. This method that optimizes a real-valued function $f(x,y)$ subject to the constraint $g(x,y)=0$ is referred to as the method of **Lagrange multipliers**

.

Example 5.4.1 Find the maximum and minimum values of $f(x,y)=x\,y$ on the ellipse $\frac{x^2}{4}+y^2=1$.

$\bigtriangledown f = y\,\mathbf{i}+x\,\mathbf{j}$ and $\bigtriangledown g=\frac{x}{2}\,\mathbf{i}+2y\,\mathbf{j}$. Hence at the critical points,

$$y=\lambda\frac{x}{2},\ x=\lambda\,2y. \quad\Rightarrow\quad y=0 \text{ or } \lambda^2=1.$$

(1) If $y=0$, then $x=0$. But P: $(0,0)$ is not on the ellipse. No good.

(2) If $\lambda=1$, then $x=2y$. $\therefore\ 2y^2=1$.

$$x=\pm\sqrt{2},\ \ y=\pm\frac{1}{\sqrt{2}}.$$

At $(\pm\sqrt{2},\pm\frac{1}{\sqrt{2}})$, $f(x,y)=1$.

(2') If $\lambda = -1$, then $x = -2y$. $\therefore\ 2y^2 = 1$.

$$x = \mp\sqrt{2}, \quad y = \pm\frac{1}{\sqrt{2}}.$$

At $(\mp\sqrt{2}, \pm\frac{1}{\sqrt{2}})$, $f(x, y) = -1$.

(3) Therefore, extreme values of f are 1 (max) and –1 (min).

Example 5.4.2 Find the maximum and minimum values of $f(x, y) = 4x + 3y$ on the circle $x^2 + y^2 = 1$.

One can find that $\lambda = \pm\frac{5}{2}$ and f takes the maximum and the minimum at $(x, y) = \pm\left(\frac{4}{5}, \frac{3}{5}\right)$. Maximum and minimum of f are ± 5.

5.5 Suggested Exercises

1. Evaluate the limit

   (a) $\displaystyle\lim_{(x,y)\to(0,0)} \frac{\sin(x^2+y^2)}{x^2+y^2}.$

   (b) $\displaystyle\lim_{(x,y)\to(0,0)} \frac{e^{x^2+y^2}-1}{x^2+y^2}.$

2. Calculate $\dfrac{\partial f}{\partial x}$ and $\dfrac{\partial f}{\partial y}$ by the definition of the partial derivative,

$$\begin{aligned}\frac{\partial f}{\partial x}(x,y) &= \lim_{h\to 0}\frac{1}{h}\{f(x+h,y)-f(x,y)\},\\ \frac{\partial f}{\partial y}(x,y) &= \lim_{h\to 0}\frac{1}{h}\{f(x,y+h)-f(x,y)\}.\end{aligned}$$

   (a) $f(x,y)=x^2y.$

   (b) $f(x,y)=e^{2x+3y}.$

3. Determine the tangent plane of $f(x,y)$ at (x_0,y_0) :

   (a) $f(x,y)=(2xy-1)^2$ at $(1,2)$.

   (b) $f(x,y)=e^{x+2y}$ at $(1,0.5)$.

4. Compute $\dfrac{\partial y}{\partial x}$ for:

   (a) $x^2-y^2-\cos xy=0.$

   (b) $x^3-y^2+2xy=0.$

   (c) $ye^x-x^2+\sin xy=0.$

5. Define the function $f:\mathbb{R}^2\to\mathbb{R}$ by

$$f(x,y)=\begin{cases} xy(x^2-y^2)/(x^2+y^2), & \text{if } (x,y)\neq(0,0),\\ 0, & \text{if } (x,y)=(0,0).\end{cases}$$

   Show that

$$\frac{\partial^2 f}{\partial y\partial x}(0,0)\neq\frac{\partial^2 f}{\partial x\partial y}(0,0).$$

6. A function $f : \mathbb{R}^2 \to \mathbb{R}$ is called **harmonic** provided that it has second-order derivatives and
$$\frac{\partial^2 f}{\partial x^2}(x,y) + \frac{\partial^2 f}{\partial y^2}(x,y) = 0 \text{ for all } (x,y) \text{ in } \mathbb{R}^2.$$
Investigate if the following function is harmonic or not.

   (a) $f(x,y) = e^x \cos y$.

   (b) $g(x,y) = x^2 - y^2$.

   (c) $f(x,y) = (x-y)^2 + (x+y)^3$.

7. Let $f(x,y,z) = \sin(xyz) + e^{x+y} + z^2 + 2$ for $(x,y,z) \in \mathbb{R}^3$. Determine $\boldsymbol{H}_{(0,0,0)}$.

8. Define $f(x,y) = x^2 + y^2 - 3xy$ for (x,y) in $\mathbb{R}^2$. Show that the function has no local extreme points.

9. Find the local maximum of $f(x,y) = e^{-\frac{1}{2}(x^2+y^2)}$.

10. Find all local maximum, local minimum and saddle points of $f(x,y) = 8x^3+6xy+y^3$.

11. Show that $(0,0)$ is a saddle point of $f(x,y) = xy$.

12. Show that the limit of function $f(x,y) = \dfrac{xy}{x^2+y^2}$ does not exists as $(x,y) \to (0,0)$.

13. Consider the function
$$f(x,y) = \begin{cases} \frac{x^2 y}{x^4+y^2}, & (x,y) \neq (0,0), \\ 0, & (x,y) = (0,0). \end{cases}$$
Is f continuous at the origin $(0,0)$?

14. Let $f(x,y) = 3x^2 - x^3 - y^2$. Verify that $(0,0)$ and $(2,0)$ are critical points of f.

15. Find the maximum and minimum values of $f(x,y,z) = x - 2y + 5z$ subject to $x^2 + y^2 + z^2 = 30$.

16. Evaluate
$$\lim_{x\to 0,\ y\to 0} \frac{x^3 y}{x^4+y^4}, \text{ and } \lim_{x\to 0,\ y\to 0} \frac{x^3 y}{x^2+y^2}.$$

17. Find the partial derivatives of $f(x,y) = \log_y x$ at $(x,y) = (e,e)$.

18. Find the maximum and minimum values of $f(x,y) = \frac{x^2}{8} + \frac{y^2}{2}$ subject to $xy = 1$.

19. Find the equation for the tangent plane at $(1,-1,3)$ on the surface of $x^2 + 2xy - y^2 + z^2 = 0.7$.

20. Find the values of $\frac{\partial f}{\partial x}$ and $\frac{\partial f}{\partial y}$ at the point $(2,-1)$ if

$$f(x,y) = x^2 + 3xy + y - 1.$$

21. Find $\frac{\partial f}{\partial x}$ and $\frac{\partial f}{\partial y}$ as functions if

$$f(x,y) = \frac{2y \sin xy}{y + \cos x}.$$

22. If $yz^4 + x^2z^3 = e^{xyz}$, find $\frac{\partial z}{\partial x}$ and $\frac{\partial z}{\partial y}$.

23. Find the extreme values of the function $f(x,y) = x^2 + y^2$ on the circle $x^2 + y^2 = 1$.

Chapter 6

Double Integration

Suppose that two continuous random variables X and Y form a joint density function $f(x, y)$ over a certain region S. One typical problem in statistics is the evaluation of the probability $P\{(X, Y) \in E\}$, where E is a subset of S. The probability can be obtained via the double integration.

6.1 Double Integral Over A Rectangular Region

Suppose that $f(x, y)$ is continuous and non-negative. We will compute the volume V of the geometric object under the curved surface of $z = f(x, y)$ over the rectangular region $E : (a, b) \times (c, d)$. See Figure 6.1. V is denoted as

$$V = \int_E \int f(x, y) \, dx \, dy$$

which reads as "the double integral of $f(x, y)$ over the region E."

The double integral is not limited to the non-negative functions, not to the rectangular shaped regions, but extends to any continuous function and to any connected subset of the plane.

Double integral is constructed by the following procedures:

(1) Divide the x-interval (a, b) into n_1 equal sub-intervals; (x_{i-1}, x_i), $i = 1, \cdots, n_1$. In the same way, divide the y-interval (c, d) into n_2 equal sub-intervals; (y_{j-1}, y_j), $j = 1, \cdots, n_2$. Hence it can be written as

$$x_i = a + i \, \Delta x, \quad y_j = c + j \, \Delta y$$

where $\Delta x = (b - a)/n_1$ and $\Delta y = (d - c)/n_2$.

(2) Over the (i,j)-th rectangular area A_{ij}, the minimum and the maximum of f are identified, written as m_{ij} and M_{ij}, respectively.

(3) Therefore, $V = \int_E \int f(x,y)\, dx\, dy$ is bounded from below by L_{n_1,n_2} and above by U_{n_1,n_2}, where

$$L_{n_1,n_2} = \sum_{i=1}^{n_1} \sum_{j=1}^{n_2} m_{ij}\, \Delta x \Delta y, \quad U_{n_1,n_2} = \sum_{i=1}^{n_1} \sum_{j=1}^{n_2} M_{ij}\, \Delta x \Delta y.$$

(4) If $\lim_{n_1 \to \infty, n_2 \to \infty} L_{n_1,n_2} = \lim_{n_1 \to \infty, n_2 \to \infty} U_{n_1,n_2}\ (= k)$, then we say that the double integral V exists and is equal to k.

Example 6.1.1 Compute $V = \int_E \int (x^2 + y)\, dx\, dy$, where $E = \{\,(x,y)\,|\,0 \le x \le 1,\, 0 \le y \le 1\}$.

Since $f(x,y) = x^2 + y$ is increasing both in x and y, $f(x,y)$'s minimum and maximum over

$$A_{ij} : x_{i-1} \le x \le x_i,\ y_{j-1} \le y \le y_j \quad (i = 1, \cdots, n_1;\ j = 1, \cdots, n_2;\ x_0 = 0,\ y_0 = 0)$$

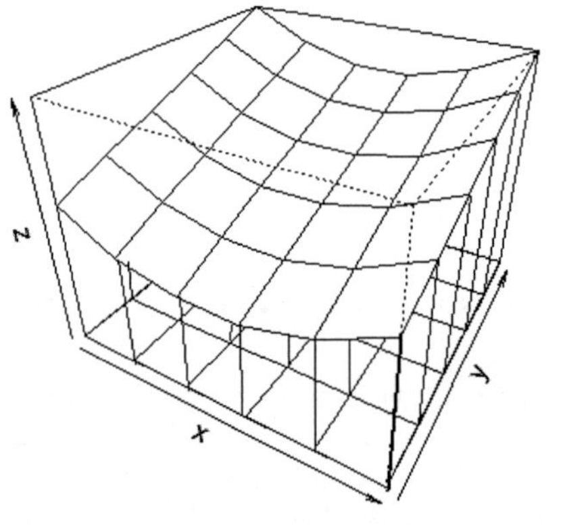

Figure 6.1: Double integral for the volume

are given by

$$m_{i,j} = x_{i-1}^2 + y_{j-1}, \quad M_{i,j} = x_i^2 + y_j.$$

Take $x_i = i/n_1$, $y_j = j/n_2$. Then

$$\begin{aligned} L_{n_1,n_2} &= \frac{1}{n_1 n_2} \sum_{i=1}^{n_1} \sum_{j=1}^{n_2} m_{ij} \\ &= \frac{1}{n_1 n_2} \sum_{i=1}^{n_1} \sum_{j=1}^{n_2} \{(i-1)^2/n_1^2 + (j-1)/n_2\} \\ &= \frac{1}{n_1^3} \cdot \frac{1}{6} n_1(n_1-1)(2n_1-1) + \frac{1}{n_2^2} \cdot \frac{1}{2} n_2(n_2-1). \end{aligned}$$

$$\begin{aligned} U_{n_1,n_2} &= \frac{1}{n_1 n_2} \sum_{i=1}^{n_1} \sum_{j=1}^{n_2} M_{ij} \\ &= \frac{1}{n_1 n_2} \sum_{i=1}^{n_1} \sum_{j=1}^{n_2} \{i^2/n_1^2 + j/n_2\} \\ &= \frac{1}{n_1^3} \cdot \frac{1}{6} n_1(n_1+1)(2n_1+1) + \frac{1}{n_2^2} \cdot \frac{1}{2} n_2(n_2+1). \end{aligned}$$

$$\therefore \quad \lim_{n_1 \to \infty, n_2 \to \infty} L_{n_1,n_2} = \lim_{n_1 \to \infty, n_2 \to \infty} U_{n_1,n_2} = \frac{1}{3} + \frac{1}{2} = \frac{5}{6}.$$

Therefore, $V = \int_E \int (x^2 + y)\, dx\, dy = \frac{5}{6}$.

The following theorem is very useful, since the double integral is obtained as repeated single-variable integrals.

Theorem 6.1.1 (Fubini's Theorem) *If $f(x,y)$ is continuous over the rectangular region $E : (a,b) \times (c,d)$, then*

$$\int_E \int f(x,y)\, dx\, dy = \int_a^b \left\{ \int_c^d f(x,y)\, dy \right\} dx = \int_c^d \left\{ \int_a^b f(x,y)\, dx \right\} dy.$$

Example 6.1.2 (Continued from Example 6.1.1) Compute

$$V = \int_E \int (x^2 + y)\, dx\, dy, \quad \text{where } E = \{(x,y) \,|\, 0 \le x, y \le 1\}.$$

Applying Fubini's Theorem

$$\begin{aligned} V &= \int_0^1 \left\{ \int_0^1 (x^2+y)\,dy \right\} dx \\ &= \int_0^1 \left[x^2 y + \frac{y^2}{2} \right]_0^1 dx \\ &= \int_0^1 \left(x^2 + \frac{1}{2} \right) dx = \frac{5}{6}. \end{aligned}$$

Alternatively,

$$\begin{aligned} V &= \int_0^1 \left\{ \int_0^1 (x^2+y)\,dx \right\} dy \\ &= \int_0^1 \left[\frac{1}{3}x^3 + x\,y \right]_0^1 dy \\ &= \int_0^1 \left(\frac{1}{3} + y \right) dy = \frac{5}{6}. \end{aligned}$$

Example 6.1.3 Compute

$$V = \int_E\!\!\int x^2\,y\,dx\,dy, \quad \text{where} \quad E = \{(x,y)\,|\,0 \le x \le 1,\, 0 \le y \le 2\}.$$

Applying Fubini's Theorem,

$$\begin{aligned} V &= \int_0^2 \left\{ \int_0^1 x^2\,y\,dx \right\} dy \\ &= \int_0^2 y \left(\int_0^1 x^2\,dx \right) dy \\ &= \left(\int_0^1 x^2\,dx \right) \left(\int_0^2 y\,dy \right) = \frac{1}{3} \cdot \frac{4}{2} = \frac{2}{3}. \end{aligned}$$

6.2 Double Integral Over A General Region

Now, we compute the double integral of $f(x,y)$ over a non-rectangular region:

$$E_1 = \{(x,y)\,|\,g_1(x) \le y \le g_2(x),\, a \le x \le b\},$$

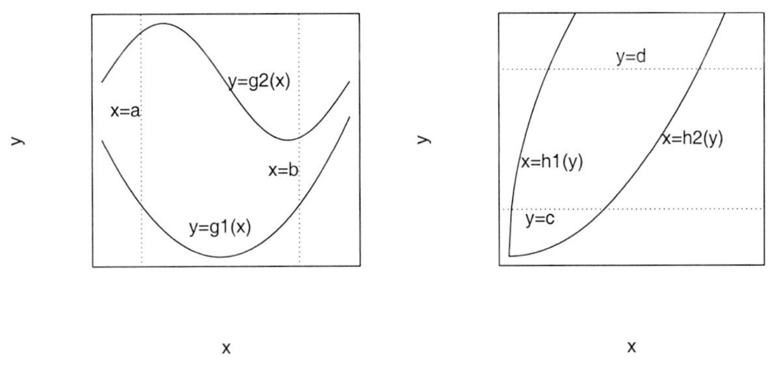

Figure 6.2: Integration regions E_1 and E_2

or

$$E_2 = \{(x, y) \mid h_1(y) \leq x \leq h_2(y),\ c \leq y \leq d\}.$$

See Figure 6.2 for the shapes of integration region.

To compute the double integral over a general region, we need to apply a strong version of Fubini's Theorem.

Theorem 6.2.1 (Fubini's Theorem, Strong Version) *Let $f(x, y)$ be continuous over the region E_1 or E_2.*

1)

$$\int_{E_1}\int f(x, y)\, dx\, dy = \int_a^b \left\{ \int_{g_1(x)}^{g_2(x)} f(x, y)\, dy \right\} dx.$$

2)

$$\int_{E_2}\int f(x, y)\, dx\, dy = \int_c^d \left\{ \int_{h_1(y)}^{h_2(y)} f(x, y)\, dx \right\} dy.$$

Example 6.2.1 Compute the volume under the plane $z = 2 - x - y$ over the region

$$E : 0 \le x \le 1,\ 0 \le y \le 1,\ x \ge y.$$

See Figure 6.3, Left:

$$\begin{aligned} V &= \int_0^1 \left\{ \int_0^x f(x,y)\,dy \right\} dx \\ &= \int_0^1 \left[\int_0^x (2 - x - y)\,dy \right] dx \\ &= \int_0^1 \left[2x - x^2 - \frac{x^2}{2} \right] dx = 1 - \frac{1}{3} - \frac{1}{6} = \frac{1}{2}. \end{aligned}$$

Alternatively (Figure 6.3, Right),

$$\begin{aligned} V &= \int_0^1 \left\{ \int_y^1 f(x,y)\,dx \right\} dy \\ &= \int_0^1 \left[\int_y^1 (2 - x - y)\,dx \right] dy \\ &= \int_0^1 \left[(2-y)(1-y) - \frac{1-y^2}{2} \right] dy = \frac{1}{2}. \end{aligned}$$

Example 6.2.2 Compute the volume under the curved surface $z = x\,y$ over

$$E : y \ge x^2,\ y \le x,\ 0 \le x \le 1,\ 0 \le y \le 1.$$

See Figure 6.4 for the integration region E.

$$\begin{aligned} V &= \int_0^1 \int_{x^2}^x f(x,y)\,dy\,dx, \quad f(x,y) = x\,y \\ &= \int_0^1 x \left(\int_{x^2}^x y\,dy \right) dx = \frac{1}{2} \cdot \left(\frac{1}{4} - \frac{1}{6} \right) = \frac{1}{24}. \end{aligned}$$

Alternatively,

$$\begin{aligned} V &= \int_0^1 \int_y^{\sqrt{y}} f(x,y)\,dy\,dx, \quad f(x,y) = x\,y \\ &= \int_0^1 y \left(\int_y^{\sqrt{y}} x\,dx \right) dy = \frac{1}{2} \cdot \left(\frac{1}{3} - \frac{1}{4} \right) = \frac{1}{24}. \end{aligned}$$

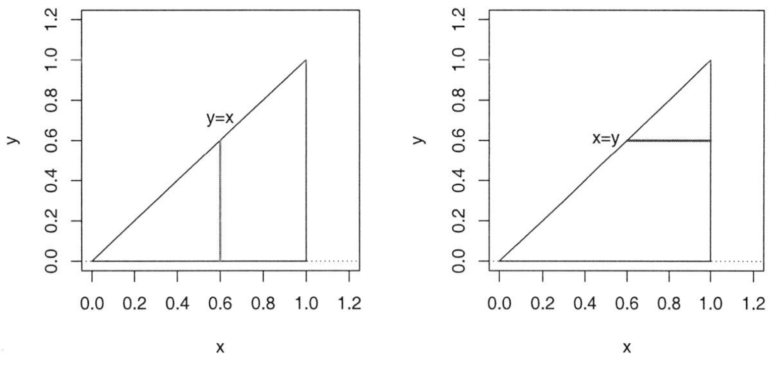

Figure 6.3: Integration region E of Example 6.2.1

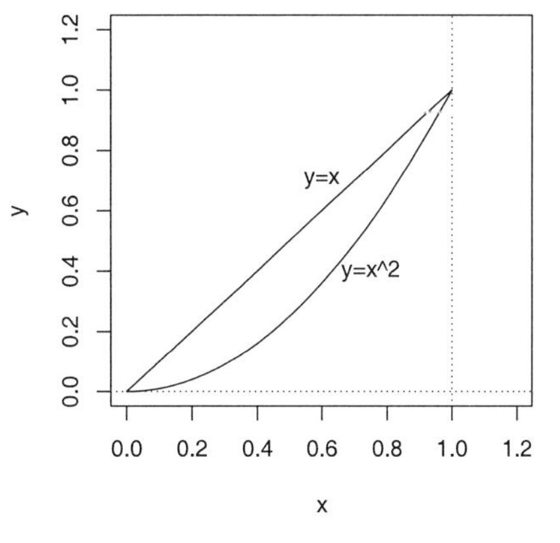

Figure 6.4: Integration region E of Example 6.2.2

Example 6.2.3 Compute the volume under the curved surface $z = \dfrac{\sin x}{x}$ over

$$E : 0 \le x \le \pi,\ 0 \le y \le \pi,\ x \ge y.$$

It is easy to compute $V_1 = \int_0^\pi \int_0^x f(x,y)\,dy\,dx$ $(= 2)$, but it is difficult to compute $V_2 = \int_0^\pi \int_y^\pi f(x,y)\,dx\,dy$. Order of the integration does matter sometimes, like in this case.

6.3 Double Integral in Polar Coordinates

The area element in orthogonal coordinates is $dx\,dy$. Now we derive the area element in polar coordinate system.

Consider a sector with the small angle $\Delta\theta$ with the radius r. The area A of the sector is $\frac{1}{2}r^2\Delta\theta$. As the radius is increased to $r + \Delta r$, the increment of the area ΔA is given by

$$\Delta A = \frac{1}{2}(r + \Delta r)^2\Delta\theta - \frac{1}{2}r^2\Delta\theta - r\,\Delta r\Delta\theta + \frac{1}{2}(\Delta r)^2\Delta\theta.$$

As $\Delta r \to dr$, $\Delta\theta \to d\theta$ (becoming much smaller),

$$dA = r\,dr\,d\theta.$$

There, $(\Delta r)^2\Delta\theta$ term disappeared since it is negligible compared to $\Delta r\Delta\theta$ term. See Figure 6.5.

Therefore, the double integral $\int_E \int f(x,y)\,dx\,dy$ can be represented as

$$\int_G \int f(r\cos\theta, r\sin\theta)\,r\,dr\,d\theta,$$

where G is the corresponding region in polar coordinates to E.

Example 6.3.1 Evaluate $\displaystyle\int_E \int e^{-x^2-y^2}\,dx\,dy$ over $E : x^2 + y^2 \leq 1$.

$$\begin{aligned}
\int_E \int e^{-x^2-y^2}\,dx\,dy &= \int_{r^2\leq 1} \int e^{-r^2}\,r\,dr\,d\theta \\
&= \int_0^{2\pi} \int_0^1 e^{-r^2}\,r\,dr\,d\theta \\
&= \int_0^{2\pi} \left[-\frac{1}{2}e^{-r^2}\right]_0^1 d\theta = \pi\left(1 - \frac{1}{e}\right).
\end{aligned}$$

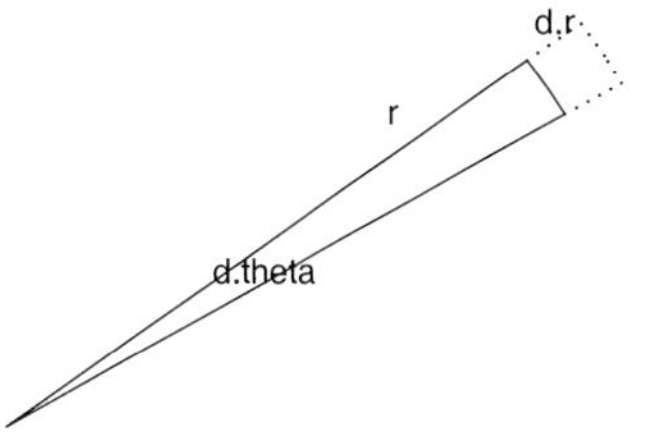

Figure 6.5: Area element in polar coordinate system

Example 6.3.2 For $E: 1 \le x^2+y^2 \le 2x$, compute $\int_E\int \sqrt{x^2+y^2}\,dx\,dy$.

Note that

$$E:\ 1 \le x^2+y^2 \le 2x \quad \Leftrightarrow \quad 1 \le r^2 \le 2r\cos\theta \quad \Leftrightarrow \quad 1 \le r \le 2\cos\theta.$$

For $1 \le 2\cos\theta$, θ should be $-\frac{\pi}{3} \le \theta \le \frac{\pi}{3}$.

$$\begin{aligned}\therefore \quad V &= \int_E\int \sqrt{x^2+y^2}\,dx\,dy \\ &= \int_{-\frac{\pi}{3}}^{\frac{\pi}{3}} \left\{\int_1^{2\cos\theta} r\cdot r\,dr\right\} d\theta \\ &= \int_{-\frac{\pi}{3}}^{\frac{\pi}{3}} \left[\frac{r^3}{3}\right]_1^{2\cos\theta} d\theta \\ &= \int_{-\frac{\pi}{3}}^{\frac{\pi}{3}} \frac{1}{3}(8\cos^3\theta - 1)\,d\theta = 2\sqrt{3} - \frac{2}{9}\pi.\end{aligned}$$

6.4 Change of Variables

In some cases, we need to transform the variables to compute the double integral $V = \int_E \int f(x,y)\,dx\,dy$. Suppose that x and y can be expressed as functions of u and v such that

$$x = x(u,v), \quad y = y(u,v) \tag{6.4.1}$$

and that the region G of (u,v) points correspond to the region E of (x,y) points. Then

$$\int_E \int f(x,y)\,dx\,dy = \int_G \int f(x(u,v), y(u,v))\,|\,J_{x,y}(u,v)\,|\,du\,dv,$$

where

$$J_{x,y}(u,v) = \det \begin{pmatrix} \frac{\partial x}{\partial u} & \frac{\partial x}{\partial v} \\ \frac{\partial y}{\partial u} & \frac{\partial y}{\partial v} \end{pmatrix} = \frac{\partial x}{\partial u}\frac{\partial y}{\partial v} - \frac{\partial x}{\partial v}\frac{\partial y}{\partial u}.$$

That is, $J_{x,y}(u,v)$ is the determinant of the 2×2 matrix composed with partial derivatives of x and y with respect to u and v, called **Jacobian**.

Remark 6.4.1 (1) We will briefly explain the reason why such a term is necessary. Consider a small rectangle with vertices

$$(u,v),\ (u+du,v),\ (u,v+dv),\ (u+du,v+dv)$$

of which the area is $du\,dv$. See Figure 6.6 [Left], in which $u+du$ is denoted by u' and $v+dv$ is denoted by v'. By the transformation of Equation (6.4.1), the parallelogram in Figure 6.6 [Right] formed in the (x,y)-plane with transformed points has the area $|J_{x,y}(u,v)\,|\,du\,dv$.

(2) Sometimes, we denote the Jacobian matrix as $J_{x,y}(u,v) = \frac{\partial(x,y)}{\partial(u,v)} = \frac{\partial(g_1,g_2)}{\partial(u,v)}$, a matrix consisting of partial derivatives of x and y with respective to u and v, respectively. That is, let

$$(x,y) = \boldsymbol{h}(u,v) : \begin{cases} x &= h_1(u,v), \\ y &= h_2(u,v); \end{cases}$$

Then, the **Jacobian matrix** , $J_{x,y}(u,v)$ is defined as

$$J_{x,y}(u,v) = \begin{pmatrix} \frac{\partial h_1}{\partial u} & \frac{\partial h_1}{\partial v} \\ \frac{\partial h_2}{\partial u} & \frac{\partial h_2}{\partial v} \end{pmatrix}.$$

and the **Jacobian** is the determinant of the Jacobian matrix,

$$\det J_{x,y}(u,v) = \frac{\partial h_1}{\partial u}\frac{\partial h_2}{\partial v} - \frac{\partial h_2}{\partial u}\frac{\partial h_1}{\partial v}.$$

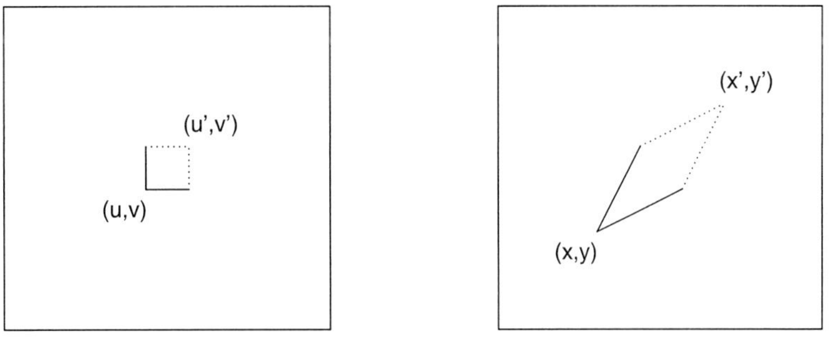

Figure 6.6: Absolute Jacobian $\mid J_{x,y}(u,v) \mid$is the ratio of two area elements

(3) Let $x = h(u)$, $a \le x \le b$, $\alpha \le u \le \beta$, and $h([\alpha,\beta]) = [a,b]$. Then,

$$\int_a^b f(x)dx = \int_\alpha^\beta f(h(u))|h'(u)|du.$$

Double integral with change of variable can be extended from the one-dimensional integral with change of variable in the following way :

Let $\boldsymbol{x} = (x,y) = \boldsymbol{h}(\boldsymbol{u})$, $\boldsymbol{u} = (u,v)$, and $\boldsymbol{h} = (h_1,h_2) : E \to G$. Then,

$$\int_E \boldsymbol{f}(\boldsymbol{x})d\boldsymbol{x} = \int_G \boldsymbol{f}(\boldsymbol{h}(\boldsymbol{u}))\left|\det[\boldsymbol{h}'(\boldsymbol{u})]\right| d\boldsymbol{u},$$

where

$$\boldsymbol{h}'(\boldsymbol{u}) = J_{x,y}(u,v) = \frac{\partial(h_1,h_2)}{\partial(u,v)} = \begin{pmatrix} \frac{\partial h_1}{\partial u} & \frac{\partial h_1}{\partial v} \\ \frac{\partial h_2}{\partial u} & \frac{\partial h_2}{\partial v} \end{pmatrix}$$

Example 6.4.1 Compute $\int_E \int \frac{2x-y}{2}\, dx\, dy$, where E is the parallelogram shown in Figure 6.7.

Let $u = 2x - y$ and $v = y$. That is, $x = \frac{u+v}{2}$, $y = v$. Hence

$$E : 2x - 2 \le y \le 2x,\ 0 \le y \le 4. \quad \Leftrightarrow \quad G : 0 \le u \le 2,\ 0 \le v \le 4.$$

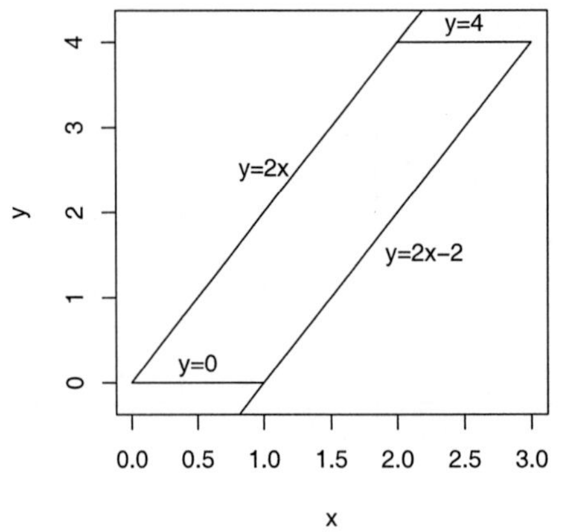

Figure 6.7: Integration region E of Example 6.4.1

Hence

$$J = \det\begin{pmatrix} \frac{1}{2} & \frac{1}{2} \\ 0 & 1 \end{pmatrix} = \frac{1}{2}.$$

$$\therefore \quad V = \int_G\int \frac{u}{2}\,|\,J\,|\,du\,dv = \int_0^2\int_0^4 \frac{u}{4}\,dv\,du = 2.$$

In the last section, we have

$$\int_E\int f(x,y)\,dx\,dy = \int_G\int f(r\cos\theta, r\sin\theta)\,r\,dr\,d\theta, \tag{6.4.2}$$

where G is the corresponding region in polar coordinates to E. We can show that, for $x = r\cos\theta$, $y = r\sin\theta$, the Jacobian is

$$J_{x,y}(r,\theta) = \det\begin{pmatrix} \cos\theta & -r\sin\theta \\ \sin\theta & r\cos\theta \end{pmatrix} = r,$$

establishing Equation (6.4.2)

Example 6.4.2 Evaluate $V = \int_{-\infty}^{\infty}\int_{-\infty}^{\infty} \frac{1}{2\pi}\, e^{-\frac{x^2+y^2}{2}}\, dx\, dy$.

$$\begin{aligned} V &= \int_{-\infty}^{\infty}\int_{-\infty}^{\infty} \frac{1}{2\pi}\, e^{-\frac{x^2+y^2}{2}}\, dx\, dy \\ &= \int_{0}^{2\pi}\left\{\int_{0}^{\infty} \frac{1}{2\pi}\, e^{-\frac{r^2}{2}}\, r\, dr\right\} d\theta \\ &= \int_{0}^{2\pi}\left[\frac{1}{2\pi}\, e^{-\frac{r^2}{2}}\right]_0^{\infty} d\theta = 1. \end{aligned}$$

Remark 6.4.2 The integral of Example 6.4.2 has the special meaning in statistics. Note that

$$\begin{aligned} 2\pi &= \int_{-\infty}^{\infty}\int_{-\infty}^{\infty} e^{-\frac{x^2+y^2}{2}}\, dx\, dy \\ &= \int_{-\infty}^{\infty}\int_{-\infty}^{\infty} e^{-\frac{x^2}{2}}\, e^{-\frac{y^2}{2}}\, dx\, dy \\ &= \int_{-\infty}^{\infty} e^{-\frac{y^2}{2}}\left\{\int_{-\infty}^{\infty} e^{-\frac{x^2}{2}}\, dx\right\} dy \\ &= \left(\int_{-\infty}^{\infty} e^{-\frac{x^2}{2}}\, dx\right)\left(\int_{-\infty}^{\infty} e^{-\frac{y^2}{2}}\, dy\right) \\ &= \left(\int_{-\infty}^{\infty} e^{-\frac{x^2}{2}}\, dx\right)^2. \end{aligned}$$

Hence

$$\int_{-\infty}^{\infty} e^{-\frac{x^2}{2}}\, dx = \sqrt{2\pi}.$$

Therefore, for the density function proportional to $e^{-\frac{x^2}{2}}$, the necessary normalizing constant should be $\frac{1}{\sqrt{2\pi}}$.

Example 6.4.3 Compute $V = \int_E \int \frac{e^{-x-y}}{x-y}\, dx\, dy$, for $1 \le x - y \le 2$, $x + y \ge 0$.

Let $s = x + y$, $t = x - y$. That is, $x = \frac{1}{2}(s+t)$, $y = \frac{1}{2}(s-t)$. E corresponds to $G : s \ge 0,\ 1 \le t \le 2$. The Jacobian is

$$J = \det\begin{pmatrix} \frac{\partial x}{\partial s} & \frac{\partial x}{\partial t} \\ \frac{\partial y}{\partial s} & \frac{\partial y}{\partial t} \end{pmatrix} = \det\begin{pmatrix} \frac{1}{2} & \frac{1}{2} \\ \frac{1}{2} & -\frac{1}{2} \end{pmatrix} = -\frac{1}{2}.$$

Hence

$$\begin{aligned} V &= \int_G \frac{e^{-s}}{t} \,|\, J \,|\, ds\, dt \\ &= \int_0^\infty \left(\int_1^2 \frac{e^{-s}}{t} \frac{1}{2}\, dt \right) ds \\ &= \frac{\log_e 2 - \log_e 1}{2} \int_0^\infty e^{-s}\, ds = \frac{\log_e 2}{2}. \end{aligned}$$

6.5 Triple Integral

Now, we extend the double integral to the triple integral which is of the form

$$\int\int_E\int f(x,y,z)dxdydz.$$

Like in double integration, Fubini's Theorem can be applied to evaluate the integral.

Example 6.5.1 Compute $\int\int_E\int (x+y^2+z^3)\, dx\, dy\, dz$ for $E: 0 \leq x, y, z \leq 1$.

$$\begin{aligned} \int\int_E\int (x+y^2+z^3)\, dx\, dy\, dz &= \int_0^1 \left(\int_0^1 \left(\int_0^1 (x+y^2+z^3)\, dx \right) dy \right) dz \\ &= \int_0^1 \left(\int_0^1 (\frac{1}{2}+y^2+z^3)\, dy \right) dz \\ &= \int_0^1 \left(\frac{1}{2}+\frac{1}{3}+z^3 \right) dz \\ &= \frac{1}{2}+\frac{1}{3}+\frac{1}{4} = \frac{13}{12}. \end{aligned}$$

Example 6.5.2 Compute $\int\int_E\int (x+y^2+z^3)\, dx\, dy\, dz$ for $E: 0 \leq x \leq y \leq z \leq 1$.

$$\begin{aligned} \int\int_E\int (x+y^2+z^3)\, dx\, dy\, dz &= \int_0^1 \left(\int_0^z \left(\int_0^y (x+y^2+z^3)\, dx \right) dy \right) dz \\ &= \int_0^1 \left(\int_0^z (\frac{y^2}{2}+y^3+y\, z^3)\, dy \right) dz \\ &= \int_0^1 \left(\frac{1}{2}z^3+\frac{1}{4}z^4+\frac{1}{2}z^5 \right) dz \\ &= \frac{1}{24}+\frac{1}{20}+\frac{1}{12} = \frac{7}{40}. \end{aligned}$$

6.6 Suggested Exercises

1. Find the integral of the function $f(x, y) = xy$ over the plane region defined by $y = e^x$, $0 \leq x \leq 1$.

2. Evaluate
$$\int\int_A e^{-y} dxdy, \quad A = \{(x, y) : x \leq y \leq 2x, \ x \geq 0\}.$$

3. Evaluate
$$\int\int_R e^{-(x+y)} dxdy, \quad \text{where } R = \{x/y \leq 2, \ 0 < x < \infty, \ 0 < y < \infty\}.$$

4. Show that
$$\lim_{n\to\infty} \int_0^1 \int_0^1 x^n y^n dxdy = 0.$$

5. Evaluate
$$\int_0^\pi \int_0^x x \sin y \, dydx.$$

6. Show that
$$\int_0^\infty \int_0^\infty \frac{dxdy}{(1 + x^2 + y^2)^2} = \frac{\pi}{4}.$$

7. Evaluate the following integral
$$\int\int_D y\sqrt{x} \, dxdy, \quad D = \{(x, y)|x > 0, \ y > x^2, \ y < 2 - x^2\}.$$

8. Evaluate
$$\int\int_R e^{-(x+y)} dxdy, \quad \text{where } R = \{x/y \leq 2, \ 0 < x < \infty, \ 0 < y < \infty\}.$$

9. Suppose that the function $f : \mathbb{R} \to \mathbb{R}^+$ has derivatives of all orders and that
$$\begin{cases} f'(x) + xf(x) = 0, & \text{for all } x, \\ f(0) = \frac{1}{\sqrt{2\pi}}. \end{cases}$$

   (a) Does $\int_0^\infty f(x)dx$ converge? If it converges, then using a double integral in a polar form evaluate $\displaystyle\int_{-\infty}^\infty f(x)dx$.

(b) Find all local maximum, local minimum and saddle points of $f(x)f(y)$, if any.

10. We would like to determine the convergence of the following improper integral:

$$\int_1^\infty \int_{\sqrt{3}x}^\infty \frac{1}{(x^2+y^2)^2}\frac{y}{x}dydx. \tag{6.6.3}$$

(b) Evaluate the integral

$$\int\int_R \frac{1}{(x^2+y^2)^2}\frac{y}{x}dydx, \quad R = \{(x,y) \ : \ x^2+y^2 \le 1\}$$

(b) Determine the convergence of the improper integral (6.6.3).

11. Show that

$$\int_0^3 \int_1^{\sqrt{4-y}} (x+y)dxdy = \int_1^2 \int_0^{4-x^2} (x+y)dydx.$$

12. Suppose that the function $f : \mathbb{R} \to \mathbb{R}$ is continuous. Prove that, for each $x \ge 0$,

$$\int_0^x (x-t)f(t)dt = \int_0^x \int_0^t f(s)dsdt.$$

13. Let $D = \{(x,y) \mid x^2+y^2 \le 1, \ x \ge 0, \ y \ge 0\}$. Evaluate

$$\int_D\int \frac{1}{2}e^{x^2+y^2}dxdy.$$

14. Evaluate

$$\int\int_A e^{-y}dxdy, \quad A = \{(x,y) : x \le y \le 2x, \ x \ge 0\}.$$

15. Evaluate

$$\int\int_R e^{-(x+y)}dxdy, \ \text{ where } R = \{x/y \le 2, \ 0 < x < \infty, \ 0 < y < \infty\}.$$

16. Let

$$f(x,y) = \begin{cases} xy, & 0 < x < 1, 0 < y < 1, \\ 0, & \text{otherwise.} \end{cases}$$

Find the volume of f between $y = x$ and $y = 0$.

17. Evaluate

$$\int_0^\pi \int_0^x x \sin y \, dydx.$$

18. Evaluate the double integral by interchanging the order of interals.

    (a) $\displaystyle\int_0^1 \int_y^1 e^{x^2} dxdy = \frac{e-1}{2}.$

    (b) $\displaystyle\int_0^2 \int_{y/2}^1 ye^{x^3} dxdy.$

19. Rewrite the following integral with the integration done first with respect to x, and then with respect to y,

$$\int_0^2 \int_{x^3}^{10-x} f(x,y)dydx.$$

20. Show that

$$\lim_{n\to\infty} \int_0^1 \int_0^2 x^n y^n dxdy = 0.$$

21. Evaluate the integral using the change of variables :

    (a) $\displaystyle\int_0^4 \int_{y/2}^{(y/2)+1} e^{\frac{2x-y}{2}} dxdy.$

    (b) $\displaystyle\int_0^3 \int_0^4 \int_{y/2}^{(y/2)+\pi/4} \cos\left(\frac{2x-y}{2}\right) + e^{\frac{z}{3}} dxdydz.$

22. Show that

$$\int_0^\infty \int_0^\infty \frac{dxdy}{(1+x^2+y^2)^2} = \frac{\pi}{4}.$$

Appendix A

Using R for Calculus

R is an open source software for statistics (`http://www.r-project.org`), which consists of the core module and hundreds of contributed modules called packages. R is well known for its functionality in statistical computing, also useful and versatile in doing mathematics as well.

A.1 Differentiation with R

First of all, you need to install `mosaic` package available at R's CRAN site and need to download the associated manual. R's `mosaic` offers symbolic calculation of derivatives. For example,

```
> library(mosaic)
> D(x^2 ~ x)
  function (x)
  2 * x
```

where `D(` $\cdot \sim$ `x)` is the `mosaic` function for the differentiation with respect to `x`. The output stands for

$$\frac{d}{dx}x^2 = 2x.$$

See Table A.1 for more examples.

Table A.1: R mosaic's symbolic differentiation

R's mosaic package	Mathematical formula

`> D(x^n ~ x)`
`function (x, n)`
`x ^ (n-1) * n`
	$\frac{d}{dx}x^n = nx^{n-1}$

`> D(exp(x) ~ x)`
`function (x, n)`
`exp(x)`
	$\frac{d}{dx}e^x = e^x$

`> D(log(x,e) ~ x)`
`function (x, e)`
`2/x`
	$\frac{d}{dx}\log_e x = \frac{1}{x}$

`> D(sin(x) ~ x)`
`function (x)`
`cos(x)`
	$\frac{d}{dx}\sin x = \cos x$

`> D(cos(x) ~ x)`
`function (x)`
`-sin(x)`
	$\frac{d}{dx}\cos x = \sin x$

`> D(tan(x) ~ x)`
`function (x)`
`1/cos(x)^2`
	$\frac{d}{dx}\tan x = \frac{1}{\cos^2 x}$

R's mosaic can handle somewhat complex functions. For

$$\frac{d}{dx}\frac{1}{\sqrt{1-x^2}} = \frac{x}{\left(\sqrt{1-x^2}\right)^3},$$

an equivalent expression can be obtained via the R mosaic script:

```
> D(1/sqrt(1-x^2) ~ x)
  function (x)
  0.5 * (2 * x * (1 - x^2)^-0.5)/sqrt(1 - x^2)^2
```

Here is another example: for

$$\frac{d}{dx}\frac{1}{\sqrt{2\pi}}\exp\left(-\frac{x^2}{2}\right) = -\frac{x}{\sqrt{2\pi}}\exp\left(-\frac{x^2}{2}\right),$$

R's mosaic shows that

```
> D(1/sqrt(2*pi)*exp(-x^2/2) ~ x)
  function (x)
  -(1/sqrt(2 * pi) * (exp(-x^2/2) * (2 * x/2)))
```

R's `mosaic` has a superb graphic utility in drawing the function and its derivative. For $f(x)$, the following R script produces the graphs of $f(x) = \frac{1}{\sqrt{2\pi}} \exp\left(-\frac{x^2}{2}\right)$, the first-order derivative $g(x)\,(= f'(x))$ and the second-order derivative $h(x)\,(= f''(x))$, for $-4 \le x \le 4$.

```
> f <- makeFun(1/sqrt(2*pi)*exp(-x^2/2) ~ x)
> plotFun(f(x) ~ x, xlim=c(-4,4), ylim=c(0,0.5), lwd=2, col="blue")
> g <- D(f(x) ~ x)
> plotFun(g(x) ~ x, xlim=c(-4,4), ylim=c(-1,1), lwd=2, col="blue")
> ladd(panel.abline(h=0,v=0,col="gray",lty="dotted"))
> h <- D(f(x) ~ x & x)
> plotFun(h(x) ~ x, xlim=c(-4,4), ylim=c(-1,1), lwd=2, col="blue")
> ladd(panel.abline(h=0,v=0,col="gray",lty="dotted"))
```

In the above R script, `makeFun(` $\cdot \sim$ `x)` defines a function of `x`, `plotFun(` $\sim$ `x)` plots a function of `x`, and `ladd()` adds graphical elements to the existing plot. See Figure A.1 and Figure A.2.

R's `mosaic` can find the roots of the given function, $f(x) = 0$, by `findZeros(` $\cdot \sim$ `x)`.

For instance, let $f(x) = (x-1)(x-2)$. The roots of $f(x) = 0$ are 1 and 2.

```
> f <- makeFun((x-a)*(x-b) ~ x, a=1, b=2)
> findZeros(f(x) ~ x, x.lim=range(-2,6))
  [1] 1.000015 1.999984
```

As a non-trivial example, let us find the roots of $h(x)\,(= f''(x))$, where $f(x)$ is the probability density of standard normal distribution, $f(x) = \frac{1}{\sqrt{2\pi}} \exp\left(-\frac{x^2}{2}\right)$. In below, `D(` $\cdot \sim$ `x & x)` requests the second-order differentiation with respect to `x`.

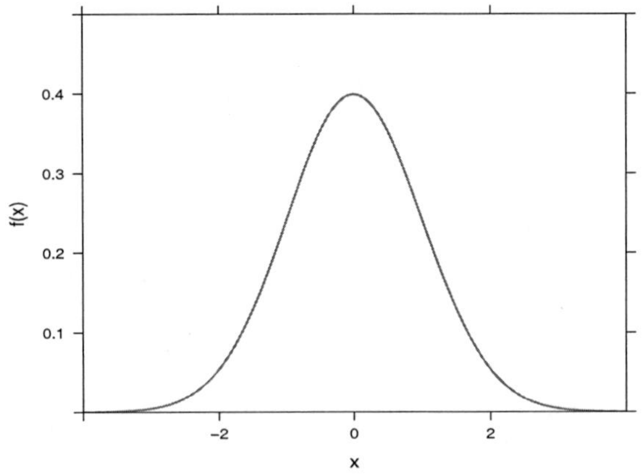

Figure A.1: The graph of $f(x) = \frac{1}{\sqrt{2\pi}} \exp\left(-\frac{x^2}{2}\right)$

```
> h <- D(f(x) ~ x & x)
> plotFun(h(x) ~ x, xlim=c(-4,4), ylim=c(-1,1), lwd=2, col="blue")
> ladd(panel.abline(h=0,v=c(-1,1),col="gray",lty="dotted",lwd=2))
> roots <- findZeros(h(x) ~ x, xlim=c(-5,5))
> round(roots,4)
 [1] -1  1
```

Thus, the probability density function $f(x) = \frac{1}{\sqrt{2\pi}} \exp\left(-\frac{x^2}{2}\right)$ has two points of inflection: -1 and 1.

Example A.1.1 Let a function f be defined on the interval $(-5, 5)$ by

$$f(x) = x^3 - 12x - 5.$$

(1) Draw the graph of $f(x)$.

(2) Find the local extremes and the points of inflection.

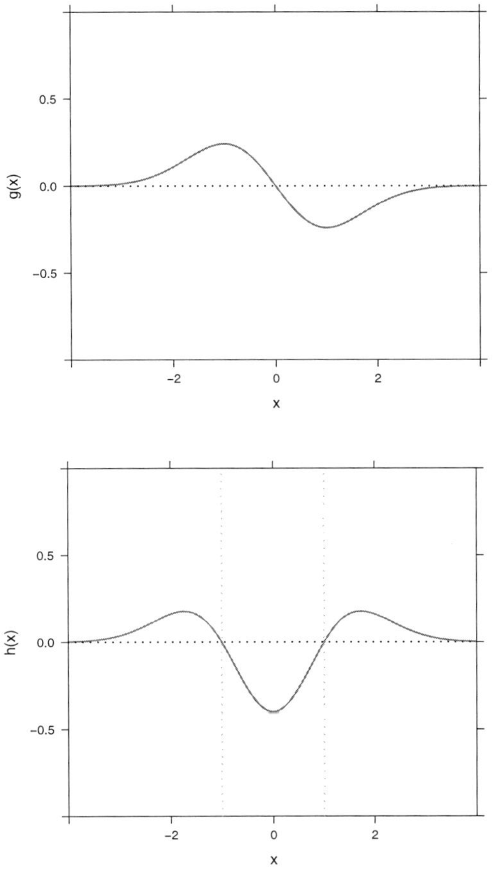

Figure A.2: The graph of $g(x)\,(=f'(x))$ and $h(x)\,(=f''(x))$

Then, we get :

Local maximum $(-2.0, 11)$, Local minimum $(2.0, -21)$, Points of inflection $(0.0, -5)$.

Example A.1.2 With $f_1(x) = \frac{1}{\sqrt{2\pi}} \exp\left(-\frac{x^2}{2}\right)$, the function $f_2(x)$ is defined as

$$f_2(x) = 0.5 f_1(x - 1.25) + 0.5 f_1(x + 1.25).$$

(1) Draw the graph of $f_2(x)$.

(2) Draw the graph of $f_2'(x)$.

(3) Draw the graph of $f_2''(x)$.

(4) Find the local maximums and minimums of $f_2(x)$.

Then, we get :

$$\text{Local maximums } (\pm 1.1, 0.21), \text{ Local minimum } (0, 0.18).$$

A.2 Integration with R

R's `mosaic` package provides a function `antiD(` $\cdot \sim$ `x)` for integrals (or anti-differentiation). For instance, let $f(x) = \sin x$, of which the anti-derivative is $F(x) = 1 - \cos x$ when a condition is given as $F(0) = 0$. Answer be obtained via the R `mosaic` script such as

```
> f <- makeFun(sin(x) ~ x)
> plotFun(f(x) ~ x, xlim =c(0,2*pi), ylim=c(-1.5,1.5))
> ladd(panel.abline(h=0,v=0,col="gray",lty="dotted",lwd=2))
> F <- antiD(f(x) ~ x)
> plotFun(F(x.to = x) ~ x, xlim =c(0,2*pi), ylim=c(0,3))
```

The result can be displayed as a graph. See Figure A.3. By plugging numbers in `F`, one obtains the definite integrals, like `F(pi)` $= 2.0$ and `F(2*pi)` $= 0.0$.

Example A.2.1 Consider $f(x) = x^{\alpha-1}e^{-x}$, $x \geq 0$ which is a Gamma function with shape parameter $\alpha\,(> 0)$. In Chapter 3, it is shown that $\int_0^\infty x^{\alpha-1}e^{-x}du = (\alpha-1)!$ for $\alpha = 1, 2, 3, \cdots$.

```
> f <- makeFun(exp(-x)*x^(alpha-1) ~ x, alpha=5)
> F <- antiD(f(x) ~ x)
> plotFun(F(x) ~ x, xlim =c(0,20))
> F(Inf)
  [1] 24
```

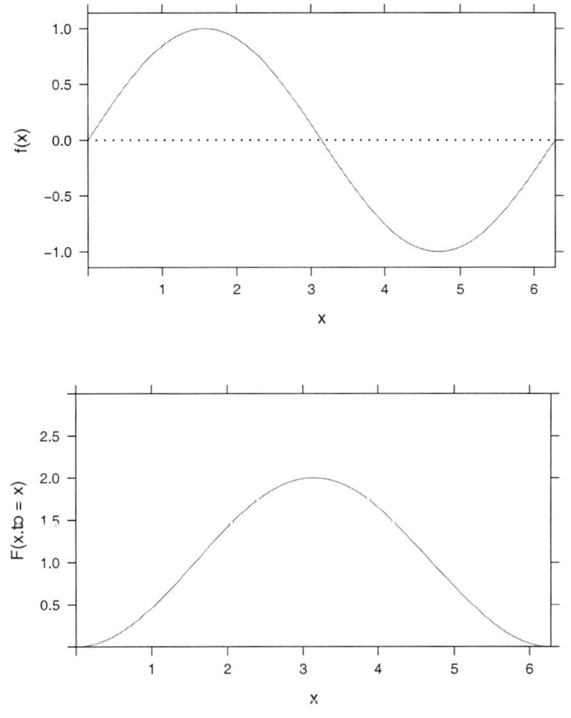

Figure A.3: The graph of $f(x) = \sin(x)$ (Above) and the graph of its anti-derivative $F(x)$ (Below)

Figure A.4 shows the graph of $f(x)$ for $\alpha = 5$ and its anti-derivative $F(x)$, where

$$F(x) = \int_0^{\infty} u^{\alpha-1} e^{-u} du, \quad u \geq 0; \quad F(0) = 0.$$

In particular, $F(\infty) = \int_0^{\infty} u^{5-1} e^{-u} du = 4!\,(= 24)$

With R's `mosaic antiD( · ~ x)` and `findZeros( · ~ x)`, one can calculate quantiles of continuous probability distributions. For instance, consider $\chi^2(n)$, the chi-square distribution with n degrees of freedom. The probability density function of $\chi^2(n)$ is given

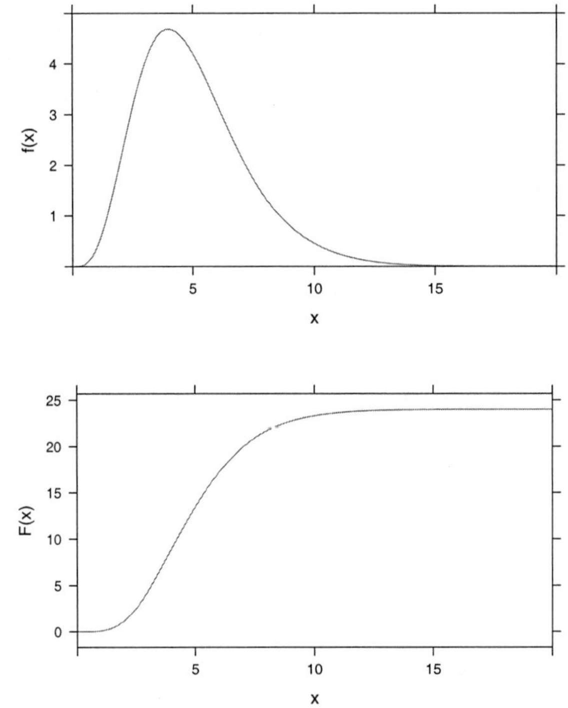

Figure A.4: The graphs of $f(x) = x^4e^{-x}$, $x > 0$ (Above) and the graph of its anti-derivative $F(x)$ (Below)

as

$$f(x) = \frac{1}{\Gamma(n/2)\, 2^{n/2}} x^{\frac{n}{2}-1} e^{-\frac{x}{2}}, \quad x \geq 0,$$

where $\Gamma(\alpha) = \int_0^\infty u^{\alpha-1}e^{-u}du$, $\alpha > 0$ by definition. To find the 0.95 quantile of $\chi^2(8)$, one may write the R `mosaic` script as follows. See Figure A.5.

```
> f <- makeFun(exp(-x/2)*x^(n/2-1)/gamma(n/2)/2^(n/2) ~ x, n=8)
> F <- antiD(f(x) ~ x)
> plotFun(F(x) ~ x, xlim =c(0,20), ylim=c(0,1.1))
```

```
> ladd(panel.abline(h=0.95,col="red",lty="dotted",lwd=2))
> findZeros(F(x)-0.95 ~ x, xlim=c(0,20))
  [1] 15.50733
```

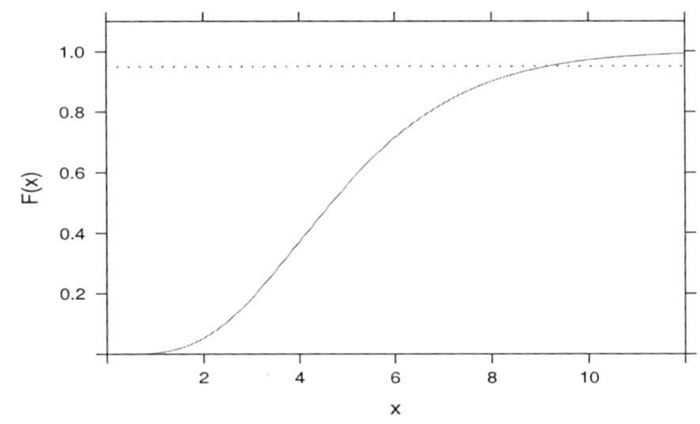

Figure A.5: The graph of $\chi^2(8)$'s distribution function $F(x)$ with the horizontal line at 0.95.

The result may be compared to `qchisq(0.95,8)` which yields 15.50731.

A.3 Partial Differentiation with R

R's `mosaic` package also provides an easy graphing of the function $f(x,y)$ of two variables x and y and its associated partial derivatives $\frac{\partial}{\partial x}f(x,y)$ and $\frac{\partial}{\partial y}f(x,y)$.

Example A.3.1 Consider

$$f(x,y;r) = \frac{1}{2*\pi*\sqrt{1-r^2}}\exp\left[-\frac{x^2+y^2-2rxy}{2(1-r^2)}\right], \tag{A.3.1}$$

a bivariate normal density with unit standard deviations and the correlation coefficient r.

```
> f <-
+       makeFun(1/(2*pi*sqrt(1-r^2))*exp(-(x^2+y^2-2*r*x*y)/2/(1-r^2))
+                     ~ x & y, r=0.5)
```

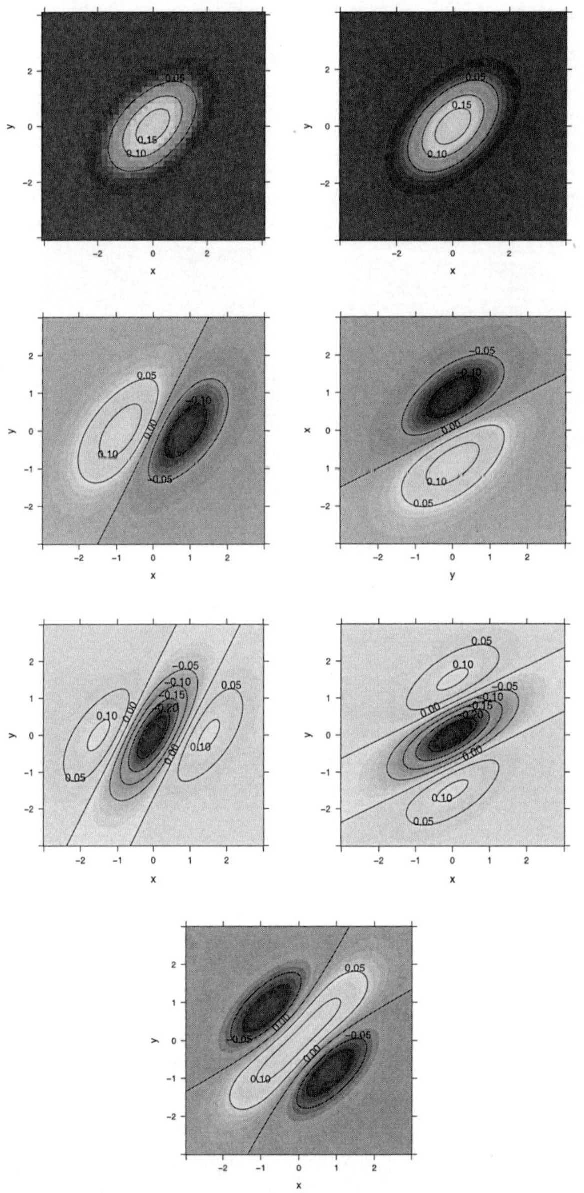

Figure A.6: The graphs of bivariate normal density $f(x,y)$ (First row), its first derivatives $\frac{\partial}{\partial x} f(x,y)$ and $\frac{\partial}{\partial y} f(x,y)$ (Second row) and its second partial derivatives $\frac{\partial^2}{\partial x^2} f(x,y)$, $\frac{\partial^2}{\partial y^2} f(x,y)$ (Third row), $\frac{\partial^2}{\partial x \partial y} f(x,y)$ (Fourth row)

```
> plotFun(f(x,y) ~ x & y, xlim=c(-3,3),ylim=c(-3,3))
> plotFun(f(x,y) ~ x & y, xlim=c(-3,3),ylim=c(-3,3),npts=300)
> g1  <- D(f(x,y) ~ x)
> g2  <- D(f(x,y) ~ y)
> plotFun(g1(x,y) ~ x & y, xlim=c(-3,3),ylim=c(-3,3),npts=300)
> plotFun(g2(x,y) ~ y & x, xlim=c(-3,3),ylim=c(-3,3),xlab="y",ylab="x", npts=300)
> h11 <- D(f(x,y) ~ x & x)
> h12 <- D(f(x,y) ~ x & y)
> h22 <- D(f(x,y) ~ y & y)
> plotFun(h11(x,y) ~ x & y, xlim=c(-3,3),ylim=c(-3,3),npts=300)
> plotFun(h22(x,y) ~ y & x, xlim=c(-3,3),ylim=c(-3,3),xlab="y",ylab="x", npts=300)
> plotFun(h12(x,y) ~ x & y, xlim=c(-3,3),ylim=c(-3,3),npts=300)
```

Figure A.7 shows the graphs of $f(x, y)$ given in Equation (A.3.1) and its associated first-order and second-order partial derivatives. In Rs `mosaic`'s graphing function `plotFun(` $\cdot \sim$ `x & y)`, `npts` option controls the number of points for each dimension to evaluate the function (with the default value 50).

A.4 Double Integration with R

It seems that R's `mosaic` package does not have a function for double integrals. However, using R's one-dimensional integration function `integrate( )` can be adapted for the double integration.

Example A.4.1 Consider $\int_0^1 \int_0^1 f(x,y)\,dxdy$, where $f(x,y) = x^2 + y^2$. We may write the R script as

```
> integrate(function(y) {
+       sapply(y, function(y){
+           integrate(function(x) x^2+y, 0, 1)$value
+           })
+ }, 0, 1)
```

The output is

```
0.833333 with absolute error < 4.3e-14
```

which is essentially equal to $\frac{5}{6}$, which is the exact value.

For the same integrand, we may even calculate the double integral over a non-rectangle area $A = \{\,(x,y) : 0 < y < x < 1\,\}$.

```
> integrate(function(y) {
+       sapply(y, function(y){
+           integrate(function(x) x^2+y, y, 1)$value
+           })
+ }, 0, 1)

0.4166667 with absolute error < 4.6e-15
```

which is comparable to $\int_0^1 \int_y^1 (x^2 + y)\,dxdy = \frac{5}{12}$.

Example A.4.2 Calculate $\displaystyle\int_0^\infty \int_0^\infty \frac{1}{2\pi} \exp\left[-\frac{x^2+y^2}{2}\right] dxdy$. The answer is 0.25.

Example A.4.3 Calculate $\displaystyle\int\!\!\int_A \sqrt{x^2+y^2}\,dxdy$, where $A = \{(x,y) : x^2 + y^2 \geq 1,\ (x-1)^2 + y^2 \leq 1\}$. The answer is $2\sqrt{3} - \frac{2}{9}\pi$ or 2.76597.

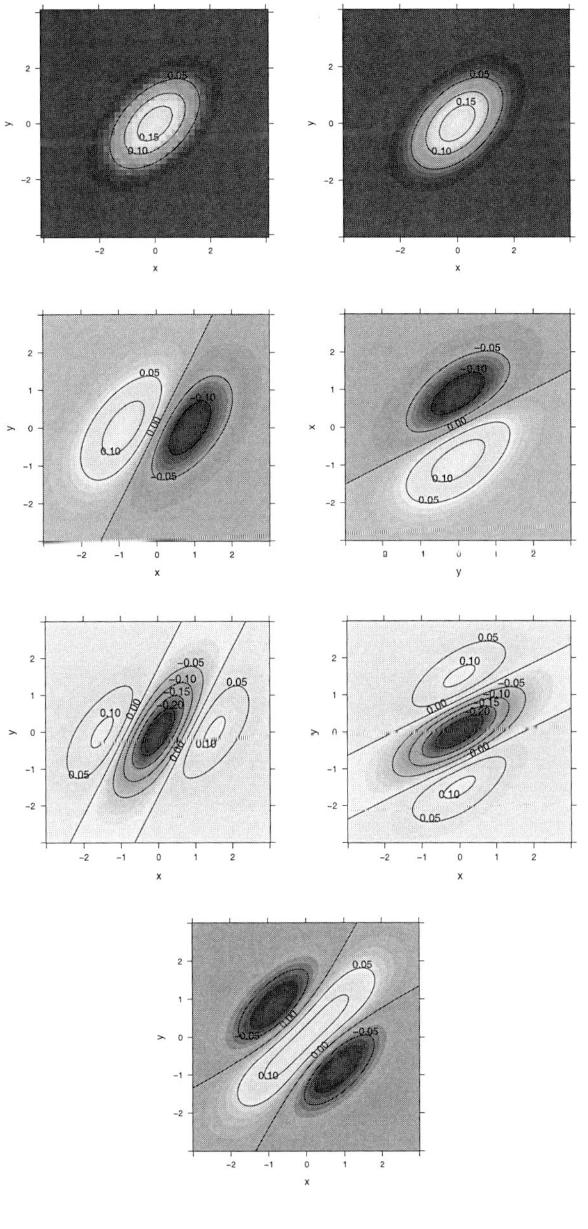

Figure A.7: The graphs of bivariate normal density $f(x,y)$ (First row), its first derivatives $\frac{\partial}{\partial x}f(x,y)$ and $\frac{\partial}{\partial y}f(x,y)$ (Second row) and its second partial derivatives $\frac{\partial^2}{\partial x^2}f(x,y)$, $\frac{\partial^2}{\partial y^2}f(x,y)$ (Third row), $\frac{\partial^2}{\partial x \partial y}f(x,y)$ (Fourth row)

Index